Gwen and the Art of Tractor Travel

Gwen and the Art of Tractor Travel

JOSEPHINE ROBERTS

Old Pond Publishing

First published 2011

ISBN 978-1-906853-84-6

A catalogue record for this book is available from the
British Library

Published by
Old Pond Publishing
Dencora Business Centre
36 White House Road
Ipswich
IP1 5LT
United Kingdom

www.oldpond.com

Cover design by Scott James
Typeset by Galleon Typesetting, Ipswich
Printed and bound in Malta by Gutenberg Press

CONTENTS

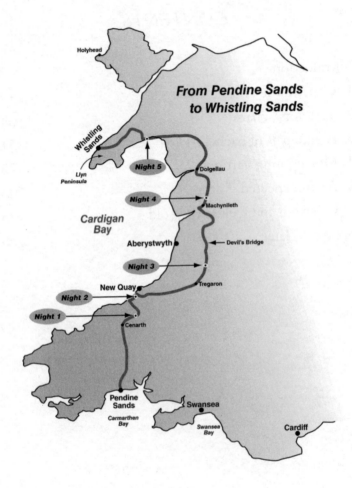

INTRODUCTION

SOME people probably think that vintage tractors are quite un-cool. Having a passion for 'old iron', as the Americans call it, may be seen as even more 'anoraky' than stamp collecting or train-spotting. But to me old tractors are beautiful. They hark back to an era when engineering was simple, elegant, easy-on-the-eye and not all covered in plastic.

Old tractors also have a distinct edge in that they keep working, pretty much forever, unlike many of today's machines, whose complicated electronics start to pack up as soon as the warranty expires. Personally, I'd rather be seen driving an early Field Marshall tractor through a busy town, than behind the wheel of a shiny, red Lamborghini.

I didn't own my first tractor until I was thirty. It just wasn't something women my age did. In a way, it just didn't sit well with other things that I thought defined me. I liked Nick Cave, for instance, and reading the *Guardian*. While the sort of people I'd seen who owned old tractors wore flat caps, boiler suits and nutter jumpers (Christmas jumpers in the style of Val Doonican). I didn't have any of these items in my wardrobe. I might not have been able to distinguish diesel from tractor vaporising oil – but I thought what the hell.

In a happy coincidence, it just so happened that I'd recently bought a ramshackle old smallholding, in a romantic bid to get back to my family's hill farming roots. Retrospectively, this probably contributed to my thinking, for the first time ever, 'I am in need of a tractor,' or maybe just, 'I want a tractor.' I'm not absolutely sure which.

I know that I wanted to use one and not just show it off the way some people do. So I also knew that I had to get one that would be capable of performing most of the tasks around a seven-acre holding. Then at least it would look like I had a proper reason to own a tractor. This factor, coupled with cash limitations, forced me to stay away from the really arcane models. You can get tractors from the 1920s and they look amazing – really more like little steam engines than tractors in fact – but you are likely to pay thousands for one and very unlikely to be able to use it to bale hay. When the inevitable occurs and it breaks down you might discover that there's only one chap alive in Britain who knows how to fix it. The odds are high that he is hiding under a flat cap in a shed somewhere on the Isle of Skye.

In the end I played it safe. My tractor came from 1959/60 – the era just before things went pear-shaped and plastic. It wasn't expensive to buy or complicated to use, and it isn't difficult or costly to maintain; at least not like the real antique machines or, for that matter, ultra-modern tractors. It's a Massey Ferguson and it's bright red.

I've had numerous adventures on my little tractor. It has made thousands of bales of hay – more than paying for itself in the first season. Only twice has it ever broken down, which was down to there being dirt in the fuel or, as the farmers say, 'shit in the tank': a comment that should never be taken as an order. My tractor has powered all sorts of machines and it has been with me to vintage rallies. On one occasion, it nearly killed me, or rather, I nearly killed myself on it by not adjusting something I should have adjusted on an implement that I was using.

You see while tractors aren't at all difficult to drive, they can be quite unforgiving. Say, for instance, that you are in the wrong gear going down a steep hill, well that can be curtains, which isn't usually the case with a car. If it were the case, we probably wouldn't see anywhere near the amount of traffic on the roads that we do.

Sometimes women think that some of the things that men do – for instance driving tractors and mixing cement – are somehow difficult but that isn't the case at all. Of course men definitely do things that are difficult but I wouldn't class driving a tractor as one of them. It's not really any harder than driving a car, a bit different perhaps, but not any harder for sure.

Now most will agree that a lot of crap is talked in pubs. Well, one day I was enjoying that particular pastime when I noticing myself saying that I'd love to drive a tractor on a holiday through Wales. Before long I was saying this almost every time I'd had a drink and I started becoming aware that I was becoming a talker, instead of a doer, which made me feel that I'd better just shut up and do it.

Then I had an unplanned pregnancy. In a way, this further spurred me on to do the trip because I knew that babies can seriously alter your life and that if I didn't do it soon, then I might never do it. Once the baby was born, I might turn into a space-wagon-and-three-way-pushchair-owning type of person and then I might be too busy ironing sleep-suits and sterilising plastic items, ever again to enjoy pointless yet pleasing pastimes. My big fear about having a baby was that it might be the end of me as a person in my own right and I might never be able to do my own thing ever again.

Well, Father-to-be didn't get where he is today by telling me what I can and can't do, so he did the decent thing with regards to my tractor trip idea and said something along the lines of 'Hmm'. The whole time I was preparing for my trip, I kept meaning to look at a map and plan a route but it just never really happened. Then, the week before, I saw a road map and noted with horror that driving from the southernmost tip of South Wales would involve driving through hugely industrial areas, like Ystradgynlais and Pontypridd: places that are deeply unsuited to tractors and tents.

In my mind the trip had to have a catchy title, like 'From Somewhere to Such and Such Place', rather like from 'Here to Eternity' but with actual place names involved. I pondered over St David's Head to Holyhead, but to be honest I wanted to finish somewhere pretty, like a beach perhaps, and by no stretch of the imagination can a port full of ferries, looking like oversized dishwashers, be described as a beach.

I scoured the map for other catchy pairs of north and south names. Eventually, I came up with 'Pendine Sands' and 'Whistling Sands', and being that these two were fairly far apart – or as far apart as is possible in a country where nothing is that far apart – and not in the least bit grim, that is what I settled on. I was going to travel from 'Pendine Sands to Whistling Sands'; all right, so I admit that it wasn't exactly 'South to North', nor was it 'Here to Eternity' for that matter, but it was the best that I could come up with. It meant starting down near Tenby and driving up to the Llyn Peninsula in the north.

If I'm honest, Pendine Sands felt like an ironic place to start a tractor trip. This beach was the place where Land Speed World Records were once made: the irony being that Pendine Sands equals 'fast', whereas tractors equal 'very slow'. I liked that contrast. It was Wrexham-born John Godfrey Parry-Thomas who set the record there in 1926, and who later died trying to defend his title in the same location. In order to really live the irony

I drove as fast as I could, which wasn't fast at all, along the aforementioned beach. I managed to get a good bit of noise and smoke out of the tractor, which always gets the adrenaline going because what it lacks in speed it certainly makes up for in sound.

As it turns out, I'm glad that I did it while I could, because I have since heard that the powers-that-be have stopped all and sundry from driving vehicles along the beach. They probably did this because people with nothing better to do felt the need to replicate the sporting glories of the past by driving (annoyingly) up and down just as I had, something quite unpleasant for everyone else using the beach.

Parry-Thomas wasn't only a superb racing car driver, it turns out that he was also a brilliant engineer who patented a vast number of inventions. The little Museum of Speed, which is situated at the edge of the beach, tells the whole story of the Land Speed Record, where 171.02 miles per hour was reached by Parry-Thomas, in 'Babs', a 27,059cc rocket-shaped car, and, as it later turned out, death trap.

A year after setting the Land Speed Record, the Welshman was back on the sands trying to defend his title, when 'Babs' skidded and rolled, and Parry-Thomas was instantly killed. He was the first driver to be killed while pursuing the Land Speed Record and he was just forty-two years old. It struck me as rather odd to discover that 'Babs' was buried in the sand where she fell and the beach was never again used for Land Speed Records. Rumour has it, that in a bizarre punishment-style ritual, 'Babs' had her seats slashed and her clocks smashed before she was buried. Whether this is true or not, I don't know.

While there was no happy ending for Parry-Thomas, there was for 'Babs' when, in the late sixties, after a bit of controversy, an ex-engineering student from Bangor, called Owen Wyn Owen, dug her up, and began slowly restoring her to her former glory. Today, 'Babs' is alive, well, and living in Wales, demonstrating the fact that while vehicles can return from the grave, humans cannot, except of course on television.

History aside, now back to me. Following a set itinerary can be tedious; it often means that you are forced to spend far too long on some things and not enough on others. The fact is that it would have been entirely possible to decide in advance exactly what route I would take and to book, or at least plan, a place to stay each night, but that would have meant knowing how far I would travel each day. But that was something I couldn't anticipate, as tractor travel was new to me.

I knew that it would taint my journey, however, if I felt that 'I have to be in Llan-bla-bla by tea-time tonight,' and it might also mean that I might have to rush past unexpected gems, just because I mustn't be late for a camping arrangement at Llansuchandsuch – where's the freedom in that? Also, let's face it, Wales is hardly the Gaza Strip, and it's not as if it is a dangerous place to wander about in, so what on earth was the need for a set route plan?

With that in mind, I decided that the best plan was to have no plan at all, except to follow my one desire, which was to stay off all main roads. This meant that I needed loads of Ordinance Survey (OS) maps, so that I could find and follow the single-track roads, avoiding the main roads and big towns like some fugitive Luddite-type figure.

Not only are main roads dangerous when you can only drive at about twelve miles an hour, because you stand a good chance of being rammed up the backside by a lunatic – some people's idea of fun, but not mine – but they are also very boring to drive. The tractor that once seemed to fairly zip along the lanes, suddenly becomes painfully slow – as opposed to 'chilled out' slow – on a main road, because the thought of a boy racer with his brains in his pants is never far from one's mind.

In any case, I wanted to see something other than road signs and fast traffic. I had driven up through Wales before; I had already seen mid-Wales through the wind-screen of a car, taking in little more than the sick verges of the trunk roads flashing by at sixty, and the dusty backsides of the other cars. That was just what I wanted to get away from. I wanted to see real life, not commuter-belt, trunk-road life.

My trip would be like a cycling holiday without the sweat and certainly without the Lycra − a cycling holiday chugging on diesel. I would see great scenery, at a perfect pace; not frustratingly slow and yet, not so fast that I'd be unable to take stock of my surroundings. Most importantly, I wouldn't be viewing the world from behind a windscreen. I would smell and feel things. There would be a real sense of 'here and now', with people pegging out washing in their gardens, herons taking off from ponds, my hand on the throttle and time to look around. The point of this journey would be the journey. I had water, food and a tent on the tractor with me and there was nothing else that I needed. I would experience the open lanes and all that they had to offer and not think of them as something to get past quickly in order to be somewhere else.

Once I started viewing my journey like this, it turned from being a drive up through Wales into a little voyage of discovery. I was looking forward to spending several days all alone, because I knew full well that being alone might not happen that much again in my future. So I took my maternity leave from my job at the funeral parlour, and asked Father-to-be if he would mind towing a trailer down to Pendine Sands, with my tractor on the back.

When it came time for me to be on my way I felt a twinge of sadness saying goodbye to Father-to-be, not because I didn't want to be on my own, but because I felt a bit guilty about leaving him to drive a noisy, slow, old Land Rover, and a huge empty trailer, all the way back up north. I suddenly saw what an effort he had made, bringing me and my pesky tractor all this way down; and how he was probably just a little bit worried about me; and how he might be hoping that I would show signs of nesting, instead of taking off with an elderly tractor and a cheap and unlikely-to-be-water-proof tent. But at the end of the day, he wanted me to be happy, and that's why he was going along with it all.

Bless him; he had made sure I had all sorts with me, just in case. I had packed a stove and gas bottle, a water container, tent, food, sleeping bag and a few pairs of big knickers. Father-to-be had then insisted that I take gallons of spare diesel (as if there were no garages in darkest Wales), more tools than I would ever know how to use, a huge orange flashing light to strap onto the roll bar of my tractor (which he made me agree to fit when-ever I went onto a main road), a first-aid kit and even a vintage-type strobe location light in case I should need to be located in the dark by a midwife in a helicopter.

As for my midwife, she just sighed lots and told me to take my NHS hospital maternity notes with me, just in case. I don't think she understood why on earth anyone who didn't have to would want to drive a tractor any-where, never mind through Wales, especially while heavily pregnant, sleeping in a tent each night, peeing in a hedge and eating crap dehydrated food. Pregnant women all over the UK tear down the motorway at ninety, stressed out, ten feet from the back bumper of the car in front, some chatting on a mobile phone as they go, and that is considered fairly normal. But driving an old tractor, when you don't have to, well that's just daft.

DAY ONE

Billowing hedges, the coracle town and some peculiar signs

As you drive out of Pendine you go up a long steep hill and when you reach the cliffs, you can look back down onto the big golden sands. Long up-hills are nice on a tractor because you can really open out and get some satisfying black smoke and loud noise issuing forth. The icing on the cake is flying sparks – seeing them coming out of your exhaust means that you're getting rid of carbon build-up, on top of creating a bit of extra excitement. It's always good to set off with a flourish because it sets you up for the journey ahead.

Coming to my first junction on the little single-track road was exciting, because I needed to apply myself to the map and work out a number of things about the roads: most importantly, which single-track roads would take me north-ish, which looked the most interesting, and which wouldn't suddenly spit me out onto a main road. I made my decision based on these factors.

Sometimes there were junctions every couple of

hundred metres and it felt very freeing to make an impulse decision, to pull back the throttle and just charge off. I had to be alone for this – had there been two of us, we would have sat at the first junction saying, 'What do you think?' to one another, and perhaps later we might snipe about bad choices the other had made. This is why travelling alone is so great – it's so 'no compromise' and as a result there's no one to blame but yourself if it goes wrong – the ultimate mode of travel for a control freak!

On this fine day in early June, my little road went on winding tightly, hedge-clad, past cottages, gardens and farms, and then it opened out onto open fields. There was more wind but that didn't matter, I had a hat and I was smiling. I felt sorry for Father-to-be on his way home north because he was having to 'do roads' in the usual way: just trying to get on with the job in hand, to get home as soon as possible on the trunk roads. How lucky I felt, being able to cavort around the countryside without a care in the world.

Because I had continued to use my tractor throughout my pregnancy to do little jobs on the holding, like carrying heavy loads about and harrowing the fields, I knew I would feel fine just pottering about the lanes on it. Most of the time it is like sitting in an armchair, but an armchair that holds you in a much better spinal position than your average couch ever would. It's true that you have to keep an eye out for potholes, because the tractor has no suspension whatsoever and even a shallow-looking dent in the road can send you right out of your seat.

So it wasn't the case that my leisurely journey left me with time to twiddle my thumbs, because the lanes were a beautiful roller-coaster, going up and down, turning this way and that, with something new around every corner. With the wind in my hair, doing all of about ten miles an hour, I had a real sensation of moving along through the air, which is something you can't experience in a regular car no matter how fast you drive it.

I had intended to drive until I felt like stopping for lunch. Then I would pull up on a grass verge, have something to eat, connect the gas bottle to the stove and brew up. But then my lane went over a cute little bridge and I came face-to-face with a pub. I hadn't expected to find a pub on any of my single-track lanes because even country pubs are never really in the country – rather they tend to be situated on the edge of trunk roads. I decided that it would have been rude not to stop at the Fox and Hounds, especially as it had placed itself in such a convenient and unusually rural location. Being both pregnant and a driver, I could be sent to hell twice for drinking too much alcohol. So I had a token half-pint and a sandwich.

The Fox and Hounds was a proper country pub, not like one of those brassy, clone pubs that you get everywhere else. I was the only customer, apart from a passing jogger and his bike-riding daughter. It was handy being able to park the tractor right outside the window because all my worldly belongings were sat in the back box, covered by an ill-fitting piece of old tarpaulin.

While the landlord disappeared into the back somewhere to make a sandwich, his wife passed through the bar with a horse's head collar and a bucket of feed, and mentioned to me that she was just off to feed the horse. When I told her that I also owned one, we struck up a gory and curious conversation about horse ailments, swapping stories of the most unpleasant and expensive-to-treat equine illnesses we had ever encountered. It was a nice little moment, that conversation, and it made a refreshing change from the many pregnancy horror stories I had endured recently.

After the pub, I decided to head towards Cenarth, the back way of course. The woody little lane followed the river for miles, which provided me with a brilliant light-shafting setting for the start of my afternoon leg. I felt a twinge when the river and I eventually had to go our separate ways, but I felt more than a moment of regret when I realised that I was going to have to do bit of main road to get into Cenarth. There was no way around it.

Since I was going to be on the main road for less than a mile I told myself there wasn't much point in trying to sort through my stuff in the back box to find the awful, flashing orange beacon that can be fitted to the top of my tractor – the one that I had promised Father-to-be that I would use. After all, I've seen loads of people on main roads with tractors that are devoid of a flashing orange light on top. In any case, Father-to-be can be just too bloody cautious sometimes. Though not so cautious as to have prevented the pregnancy, I might add.

Although I had sold out by using a main road, I tried to console myself with the thought that this road, while it might not look like much today with its Mitsubishi Spacewagons and yellow Chevrons, might actually be a really ancient and historic route. I could be following an age-old trail or something like that.

Cenarth is famous for coracles, the tiny rounded boats that are made of things like willow, split ash, tar and animal skins. I like to think of the coracle as a quirky little symbol of Wales – much like harps, spinning wheels and ladies in funny hats – but in truth, coracles are found, in varying forms, all over the world: wherever anyone felt the need to make an easily carried craft from whatever materials were available. The pictures I had seen of Cenarth had shown the river as being awash with men in lovely little bowl-shaped coracles. However, that must have been an occasional occurrence, as opposed to an everyday event, because today there wasn't a single one in sight, which was a little disappointing. The beautiful bridge with its circular tunnels above the arches was still there, of course, and so was the pub with the coracle hanging over the door.

I left the tractor by the river, deciding that Cenarth didn't look like the sort of place where people stole diesel and gas bottles off the backs of lonely tractors, and went for a wander. As I approached the National Coracle Centre building, I saw a chap walking towards me wearing a wetsuit. By way of making conversation, I asked him if this was the way to the coracle centre, even though I knew full well that it was.

He told me that, yes it was, but that it was closed at the moment.

I asked him if he was a 'coracler'. He looked at me, then looked down at his trendy wetsuit and explained in an overly patient way, that no, he was a canoeist.

As he walked away, it occurred to me that while he might not have looked like a coracle owner, I probably did look like a bag lady. My hair was standing on end, I was wearing far too many coats and scarves for the time of year and I was shouldering a pair of saddlebags which were brimming with string and bits of paper. I also have a limp – the result of a rather violent argument with a VW Golf back in 1989 – which only served to add to the whole effect. I went back to the tractor for a sit down, as all of my tractor-driving clothes were weighing me down somewhat and, suddenly, I felt quite wiped out. So there I sat, studying the map. After a minute, I fired up the tractor and headed north on the smallest, wiggliest, steepest lane that I could find, vowing that one day I would return to Cenarth to investigate the intriguing sounding National Coracle Centre.

Chugging upwards, I came to the loveliest little farm: lived-in, wholly original and un-restored. I would have loved to have chatted to whoever lived here, but there didn't seem to be anyone about, except a few tourist-eating, yellow-eyed, shifty-looking sheepdogs. There was a huge hand-painted sign in the yard, reading 'Hitler Banned Hunting'. It had the look of something that had been written in a hurry or in anger perhaps. I pondered over this unexpected and perplexing statement for a while and decided it must be a reference to the fact that Blair was about to ban hunting. As I sat there I remembered that I had once read that Hitler had been a vegetarian. One might expect a vegetarian and an anti-hunting campaigner to be a benevolent sort of person, which just goes to show how wrong you can be about people.

Moving on from bizarre political slogans, I followed my twee little bluebell-flanked lane and drank in the beauty of the countryside: the winding lanes with grass up the middle, the farmhouses nestled in the folds of the rolling landscape, the wonky wooden gates, the billowing hedges, the big blue sky and, most importantly of all, the lack of racing traffic. This glorious setting made me feel as though I was driving through some real-life 'chocolate box' territory. The pace of the tractor, and the height of it, meant that I could peer over hedges, down into people's gardens, or over the open fields and out to the uncluttered horizon. I realised how much we miss seeing when we are sweeping by in our cars, all shut-in and low-down.

Of course, it was easy for me at the time to rejoice in being in the open air, because I was experiencing that most wonderful and elusive of things – a perfect sunny day. Had it been raining, I might have looked at things a bit differently. The rain issue had certainly crossed my mind. Since my lovely and primitive little tractor has no cab, the only thing I could do, in the event of rainfall, would be to don my oversized and ancient waterproofs, tie down the tarp on the back box and soldier on. I knew from many bitter experiences of riding miles in the pouring rain on horseback that after a certain amount of time the rain begins to seep. It leaks down your neck, then down the back of your pants, and you get gradually colder and colder, until it's definitely not fun any longer and you are just sitting there with your body gradually seizing up and shutting down. To further salt the wound, it wouldn't be much fun trying to get warm and dry inside a clammy, little £39.99 two-midget tent either. So, like everyone else who ever does anything outdoors in Britain, I just had to hope that it wouldn't start raining.

The tiny lanes of Wales are like a maze, except that on the lanes there are numerous different right ways to get to where you are heading. Apart from dead-ends, you can't really go wrong and, on the upside, even dead-end roads can take you to fascinating places. You might find a long, old lane that goes on for miles, or you might end up with more junctions than you can shake a map at, sometimes as many as one every few metres. With my bad memory for long sets of road instructions, I sometimes had to drive with the OS map on my lap. When I was able to take one of the rarer long and uninterrupted lanes, I could breathe a sigh of relief, knowing that I had time to relax, put the map away in its place – under my bum – and concentrate on my actual surroundings, as opposed to the paper equivalent.

My diesel-breathing, little red dragon was, as they say in Wales, 'running like a watch'. Diesels love being hot, so chugging away a steady pace all day suits them down to the ground. All was well in tractor world.

With deep-vein thrombosis and its probable causes in mind, I stopped on a grass verge to stretch my legs and, as I was walking around the tractor marvelling at how reliable it was being, I happened to touch the front tyre. I nearly burnt my hand – it was really, really hot. In alarm I thought, what if tractor tyres aren't meant to run on tarmac all day in the baking sun? And what if my tyres just, you know, melted away? I quickly put the thought out of my head, telling myself that there was no way I was going down the avenue of stopping for an hour, every two hours, to cool the tyres down – that would be more distressing than enduring a complete and sudden melt-down.

Deep down, I was half expecting, and fully dreading, that something might go wrong with the tractor. I had tools with me – more tools than, sadly, I knew what to do with. I knew I could always get advice over a phone and I didn't mind asking at a farm for help, or paying a local tractor mechanic to come out and fix it up. The only thing I really didn't want to happen was to find myself in a position where I would have to be rescued by a member of my family. That would be terrible. For one thing, I would never hear the end of it, and for another, I would feel like a failure and a fool.

You know how it would look, the kid-sister-who-accidently-got-up-the-duff goes off on some hare-brained mission and when it all goes wrong she bursts into tears, shouts for help and gets rescued by her long-suffering big brother, who is not at all amused at having to drive all day with a very slow vehicle and trailer, scouring the back lanes of Wales, looking for a broken tractor and a sobbing sister – not good. Still, Massey Ferguson 35s are the best tractors ever so a total breakdown wasn't likely to be on the agenda and, therefore, there was no point in dwelling on what could go wrong.

Later on, that first afternoon, I began to feel a sudden tiredness come over me. The feeling that you could just lie down and sleep, no matter where you are or what's going on, is apparently quite common during pregnancy. I had spotted a campsite on the OS map a couple of hours earlier and – as it seemed to be in the middle of nowhere – I had vaguely headed towards it. By now, it was probably only a few miles away, so it seemed like a convenient time to call it a day and pitch up.

I couldn't wait to set up camp; I was already dreaming about a pan full of super noodles and the cosy, plasticky smell of my tent. Then I arrived at the entrance to the campsite. Far from being what I expected – a little farm with a few tents in the back field and a place to pee in an old cowshed – this was, judging by the sign, a country-club style all singing all dancing caravan park.

Having come this far, and as there wasn't another campsite on the map for many, many miles, I thought I had better just get over my disappointment and go right in. Driving between rows of static caravans with families playing ball games outside, I felt woefully out of place. I expected to be mistaken for a handyman at any moment and be beckoned over by some caravaners wanting their bins emptied. The place had a swimming pool, an adventure playground and a bar with a sign outside which read 'Karaoke Tonite'. There wasn't another tent in sight, never mind a tractor.

Suddenly, I was inspired to take my chances in the open countryside so I turned around and chugged out, clanking loudly over the speed bumps as I went. It's not that I'm a snob – in fact quite the opposite. It wasn't that the place wasn't good enough for me; it was actually too good, too public with too much laid on. I just wanted to stay away from all that. In any case, my stuff wasn't that secure in the transport box, with only my old tarp tied over it, and I decided that I would feel a whole lot safer in a place with fewer people around. My anti-social streak showing itself, perhaps, or maybe I was just dreading that some creepy weirdo would come and try to make friends with me, grovelling under the misapprehension that I was somehow in need of his/her company. People like that really don't take hints, meaning you soon realise that you are going to have to tell them to *ffwcia 'i o 'ma*. Telling someone where to go always leaves a nasty taste in my mouth.

As I drove on up the road, away from the warm prospect of my tea and tent, there was a nip in the air, the sky wasn't as bright as it had been and I felt an unpleasant little twinge of anxiety. Here I was, miles away from home on my tractor, and I felt cold, tired, extremely pregnant and starving hungry. Although it had been a lovely day, it had also been a long one. I just felt I needed to stop. I donned my fireproof balaclava (another thing which Father-to-be had insisted I take) for some extra warmth and cheered myself up with the thought that perhaps the people who made the OS map didn't mark every existing campsite. There had to be other, smaller, lesser-known campsites: quiet places with just a tap and a toilet.

I decided to ask at the next place I passed. But that's people for you – you just can't get away from them most of the time, but when you need some, there aren't any to be found! Every house I passed looked totally deserted. Then I came to house which had a car parked outside. As I stood there knocking loudly at the wide-open backdoor I could hear rock music playing loudly from an upstairs room. I called out a few times, then gave up and went away, deciding that there must be two teenagers having frantic sex in a bedroom upstairs and the last thing they would want to do would be to discuss camping with a balaclava-clad, fur-coat-wearing pregnant tractor driver.

Pressing on, I came to a sign saying 'Museum of Power'. This was to be the second sign that day over which I would ponder the meaning. I decided to go, see for myself what a museum of power was and, moreover, to ask the occupants if they knew of a nearby campsite. So, removing the balaclava first, into the yard I drove. A man in a boiler suit walked over and waved a large adjustable spanner at me. I pulled the stop button on the tractor, because it is impossible to hold a polite conversation over the sound of a tractor running, and said hello. I explained my predicament: that I was driving up through Wales and blah blah blah. The man took it all in his stride, not surprisingly, because I later found out that I was talking to a chap who in his time had frequently driven huge, slow, vintage road rollers quite some distances. My tractor trip probably sounded like child's play to him.

Paul was his name, and his wife was Hazel, and they

were just the nicest people you could hope to bump into. They explained that they couldn't think of a campsite nearby and that the nearest ones would be miles away in the touristy areas down by the coast. Then Hazel said, 'If you like you could just camp here in the field.' To which I replied, 'That would be brilliant!' Thank heavens! I thought. Finally, I would have my pasta in a packet after all and I would have my cosy, plastic-smelling nest, right here and now, what bliss!

All that had to be done was to move the present occupant out of the field before moving me in. The occupant was a small, sharp-horned black sheep, which hailed from a remote part of the Scottish Isles and had the look of the devil about it. I was most impressed with the sheep and said that I was quite willing to spend the night with it – of course I meant this in a platonic way, I might be Welsh but there are limits – but it was not to be. Hazel told me that it liked to rip things like tents with its sharp, little horns, so it had to go into another field.

Then Hazel suggested that I come into the house for a cup of tea or to cook my food. I explained to her that I really wanted to set all my stuff up – the gas, the stove and the tent – and just play house because it was my first night and it would feel like cheating if I went in someone's house to do it all. I also felt that I had accepted enough of their kindness, without spending the evening in their home as well. So I got the tent up and my bed laid out, all while the ten-minute pasta dish was boiling away in the saucepan. Then I sat on my fold-up chair and ate my food, while surveying my kingdom and the sinking sun, feeling at one with the world and utterly contented with my lot.

Although my wonky knee joint and huge baby bump made manoeuvring about in the very low tent a little bit difficult, I slept really well that first night, as one does when one has been out riding the breeze all day. One of the many annoying downsides of being heavily pregnant, though, is that your body doesn't seem to be able to hold as much pee as it normally would do; this meant that I had to get up, find my boots, extract myself from the sleeping bag and tent and go outside several times in the middle of the night. The first time I did this there was just enough moonlight to see by and I noticed that some ten metres away from the tent was what appeared to be the shape of a figure, hunched over. It crossed my mind that it might be the black beast of Wales – the one we read about in the papers – waiting to pounce, but I was just too tired to contemplate the idea thoroughly. I returned to my nest, making sure I had my penknife next to my pillow, and went back to sleep. After all, a hungry, feral panther was no real match for a pregnant woman with a wonky knee and a tiny blunt penknife, was it?

DAY TWO

A horse trainer and a foggy Llangrannog

THE morning had arrived before I crawled out into it: the air was fresh, the sky as clear as a whistle and the day full of promise. I felt mentally great, but physically I was aching all over. I knew it wasn't the tractor driving – because that had actually been quite comfortable. Instead, it was the field that had done me in. Little did I realise that when I bought my £2.99 carry mat I could have saved myself at least £2 by just buying a newspaper. A carry mat, or at least, a cheap carry mat, is about as much good to lie on as a few sheets of paper. While the beauty of the newspaper is that it has other uses too, a carry mat has no other real purpose at all, other than to sit rolled up under your rucksack helping you to look the part when you go camping.

With a knee that can't be used for kneeling on and a gut like that of a lorry-driving darts player, I stuffed my way awkwardly out through the tiny doorway of the tent. Tea, toast and peanut butter worked wonders and after a few yawns and stretches, I was raring to go again.

Peanut butter: one of my favourite foods, but yet another thing you aren't supposed to 'do' while pregnant, along with cat poo, excessive caffeine, pregnant sheep, heavy lifting, alcohol, cigarettes, soft cheese. . . .

Gazing around, I was pleased to note that the hunched figure from the night before was in fact a clump of waist-high nettles. Before I set off, Paul offered to show me around the museum, which consisted of two large sheds, containing all sorts of big black engines. I have seen stationary engines at shows, sat there, putt-putting away and powering nothing at all and, to be honest, it's hard to tell what exactly their purpose is. At the Museum of Power, however, most of the machines were put in the setting they would have originally been in and it helped that I had a guide. Paul was able to tell me the story of each engine, what it would have been used for and how exactly it ended up being rescued and brought here.

I have never really thought that I was someone who was interested in engines, but Paul's enthusiasm for the history of these sources of power was strangely infectious. Some of these engines, in the days before electricity, would have provided the power for stately homes, breweries and mines. Outside, dotted around the place, were various road rollers (machines capable of travelling at even slower speeds than my tractor) and a colossal, grey engine that had come out of a ship. Paul, whose brainchild was this little museum, was obviously a guy who was pretty obsessed with collecting engines and I thought it was great that he had managed to tailor a business around his actual hobby. It made me think that there was a lesson to be learnt here: we should all strive to make a living doing something we enjoy because life is too short and precious to waste on an unsatisfying, stressful or just plain boring career.

When it came to saying goodbye to Paul and Hazel, I felt that my faith in human nature was well and truly restored – these people had allowed a complete stranger to camp in their field, for no gain other than the pleasure of having helped a fellow human, and that was brilliant. The media would have us believe that we are living in a dangerous world where it's each man for himself; where so many people's motto in life is, 'You can't be too careful, can you?' Well I think that you can be too careful – sometimes to the point where you can become quite paranoid, believing that there's danger around every corner, and that's no way to live.

Once again the sun was shining on me, God was in his heaven and I was in mine, and with that feeling in mind, I decided to head towards the coast. The map showed that by mid-morning I would be pretty close to where an acquaintance of my brother happened to live. This bloke was someone my brother had met on that most curious of activities – the ploughing circuit.

Now, you might think that ploughing is something that a farmer does when he wants to cultivate the land before planting a crop but there's a whole other kind of ploughing too, namely competitive ploughing. Yes, strange as it may seem, there's a ploughing set consisting of people who travel the length and breadth of Britain for no other reason than to compete in ploughing competitions. Some people plough with tractors, anything from the ancient to the ultra-modern machines, and some plough in the age-old way, using a pair of horses.

It was through horse-ploughing that my brother had met Jim and, although I had never met him myself, I had been told that it would be okay to drop by if I was passing. I was curious to see Jim and his horses, as he is well known on the circuit for his non-aggressive horse-training techniques. I suppose he is what some people would call a horse whisperer, which always sounds like the job title of some crazy American animal therapist. My brother Dafydd had been fascinated by the bond that Jim has with his horses and the way they had happily responded to his every quiet command. This is what I hoped to see on my visit.

It wasn't a detour as such, since I wanted to look at the sea and it happened that Jim's place wasn't far from the coast. I wasn't sure exactly where Jim lived but that didn't matter because I thought I would stop and ask once I got nearby. As luck would have it, Jim's neighbourhood consisted of winding, woody lanes, so it was right up my street. As I passed a cottage on the side of the road, an old man stood in the doorway open-mouthed, staring at me. He lifted his hand slowly after I waved at him. I don't think that he had never seen a tractor before; it was more likely that he'd never seen anyone that he didn't recognise driving one.

I paused at one of those little stores that open for two hours a day, except on days with a 'u' in the name. You know, one of the stores that only the locals know which two hours it will be open for. But the sign saying 'open' is always there, no matter what time of night or day it is. I fancied stopping for a can of pop but clearly I had picked the wrong two hours in which to pass. The door was locked, so I pressed on.

Later, on a narrow lane, I rounded a corner and came face-to-face with a chap long-reining a wild-eyed little pony. I backed up into a lay-by and switched off the tractor. Amazingly I had managed to come face-to-face with the man himself. Initially, I didn't realise this was him because when someone is known to be a keen ploughman and goes by the name of Cardigan Jim, you would be forgiven, I hope, for expecting him to look like a white-haired, ole timer in a pair of Pa-Walton-style dungarees and a cardigan with brown, football-shaped buttons.

This chap, however, was tall and slim and had a long flowing head of locks. He would have looked com-pletely at home with an electric guitar in his hands, which just goes to show that you shouldn't generalise about people. Something that I should know because

I've got a friend who's a lady truck driver and has the sort of polished nail extensions that you might expect to see on a beauty therapist. Plus, I have another friend who is a young plough-woman who wears shell-suits and baseball caps and would look quite at home hanging around and looking menacing in the stairwell of a sink estate. Not to mention the gas welder I once met, who was wearing a pearl necklace underneath his oily boiler suit. But that's another story.

After a cup of tea in his kitchen, Jim showed me a video. It was a film of a recent performance he had given. He and his Shire horse give a dance-like perform-ance where they pirouette around each other, bow to each other and what not. It was intriguing stuff and of particular interest to a fellow horse owner like myself because horse owners know that horses don't generally lend themselves to remembering complicated dance moves in front of large, flood-lit audiences. I decided it must be quite a special horse, with quite a special trainer too.

After the video, I got to see a live performance with the Shire horse, Tudor, performing loose in Jim's arena. There were no ropes or whips but Jim was able to direct Tudor away from him and towards him just by the use of small hand signals. The horse would also copy Jim's movements: when he did a high-stepping walk, the horse did the same alongside him and when Jim pirouet-ted, the horse pirouetted too. It was as if the horse was enjoying playing an amusing game, which was interesting because despite a lifetime of being around equines, I have never before seen much evidence to suggest that horses actually have a sense of humour. Later, we took Tudor back to the field with the other horses, which was nice to see because it meant that he led an ordinary life and wasn't some lonely, messed up, circus-performing beast.

I chatted with Jim for several hours, about horses mainly, but then the conversation got around to why exactly I was making this tractor trip. I told him all the stuff about how, very soon, I was going to have to be a responsible mother, of sorts, and that I would have to dedicate loads of what had previously been my time, to looking after the baby. He listened carefully, and then he asked me if perhaps I was running away from something? The idea had never entered my head before and hearing it put like that made me laugh out loud. I just thought: if you were running away from something, you wouldn't pick a bloody vintage tractor as a getaway vehicle would you? I like to think that if I had been running away I'd have made a better job of it than this; I'd have been on an aeroplane, heading for Barbados, possibly armed with a full set of adoption papers. . . .

Still it was an interesting thought and one that I pondered as I drove down the lane, following my pleasant morning with Jim. The thing was, I'd gone past the stage of running away from being pregnant – being terrified had happened to me at the very beginning – and now, well, I was just trying to make the most of the free time I had left.

I know that many people will think this sounds ridiculously melodramatic but I had this analogy in my mind when I first found out that I was pregnant. I thought, imagine if someone asked you to look after a creature you weren't familiar with, say a crocodile, and they wanted you to look after it for *the rest of your life*. Oh yes. Now, you are quite scared of crocodiles but you like them, from a safe distance, and you certainly wouldn't want to harm one but, crikey, you don't know anything about crocodiles and you definitely never planned on having one of your own. On top of that, the worst part of the deal is that if you don't have the crocodile, it dies.

Given the alternatives, after a lot of awful soul searching, you decide to have the crocodile and to try your best to make a decent job of looking after it. It will mean big changes, especially in the first few years, when the crocodile will need round-the-clock care and, of course, there's the crocodile's accommodation to sort out, plus food. Not to mention that you'll have to read up on crocodile care and probably attend a course of some kind, and kiss goodbye to your career. . . . Life would never be the same again but I had mostly accepted that now and I was just having a little holiday because in a few more weeks the crocodile would be arriving.

The day was still in its prime and as I felt like I'd already achieved a lot, I decided to head to the seaside, park up, get the gas stove out and boil some beans. Referring to the map, it appeared that the nearest beach which didn't involve main roads to get there, was at Llangrannog, so I decided on that. The season down here was much more advanced than at home and all the farmers were busy making silage while the sun was shining. As a result, I met several giant, glass-clad modern tractors on the lanes. I waved at them all of course and they waved back but you couldn't really see much of anyone's face, inside those high-up glass cabs. Being in a tractor like that is more like being in a lorry than anything else. I know modern tractors have many enticing attributes, but I wouldn't want to go on holiday with one, that's for sure.

The little seaside village of Llangrannog was once the home of a multi-talented, nineteenth-century woman called Sarah Jane Rees. Rees was a navigational instructor who was qualified as a Master Mariner, having been to sea, where she served as a deckhand on her father's ships. She was also a poet who won the chair at the National Eisteddfod under her bardic name of Cranogwen. Somehow, she also found time to edit a women's magazine, teach music, become a preacher and be an active member of the Temperance Movement.

I read some of this information about Rees in a fascinating paperback book, which I'd found in a charity shop back home. The book is an excellent, vast work, covering a huge array of facts and fables, all of which serve to give a vivid picture of what Wales is all about. *The Matter of Wales*, as it is called, is written by Welsh author, Jan Morris, and it manages to provide bite-sized pieces of information about all sorts of places – just enough information to keep you interested but not so much that you feel like you're reading a dry, old text-book. It is a book that you can just dip into again and again and it's hard to imagine why someone would have given it away to a charity shop. Still, I'm glad they did, or else I might never have found it.

My fascination with this book really serves to show that, like many people who have lived here all their lives, I know very little about the history and legends of the country that I call home. Wales seems to like to keep quiet about its huge history and wonderful legends. You have to bother to go out and find information about Welshness. Even many first language Welsh people are relatively clueless about the facts and fables of their country. Perhaps the Welsh are just a quiet bunch, maybe they like keep their culture locked away, only bringing it out on special occasions like their best china.

I wondered about all this as I descended carefully downhill towards Llangrannog. The village is in a shel-tered little cove and the lane drops very sharply down into it, and then climbs steeply back up the other side. You have to watch tractors on steep hills, even if you aren't pregnant. You can't just sail down in a high gear with your foot on the brake, as you might be able to do in a car. If you drove a tractor down a hill in that way the engine would rev up to fight the pressure of the brakes, and if you couldn't hold back the engine it could win and you would lose control and go over on the first sharp corner. Rolling a tractor is something which should be avoided at all costs, largely because you stand a good chance of being crushed.

The way that you manage a steep descent is to select a low gear at the top of the hill, one slow enough that you will easily be able to pull the tractor up from if you need to stop. Though sometimes, you can't tell in advance how steep a hill is going to be. The one at Llangrannog is fairly steep but as my house is on a steep hill, I managed it fine, and I was looking forward to revving the balls off the tractor, in a totally unladylike manner, on the way back up.

Although the sky was clear blue everywhere else, Llangrannog itself was completely engulfed in some seriously thick sea mist, or pea soup, as they say. Sarah Jane Rees would no doubt have known many a mist like this. I put my lights on, just in case a hurrying car failed to see me. The mist made everything silent; it even dampened the sound of the tractor. Each house I passed came at me out of the fog and it gave the village a quiet, mysterious air which was eerie in an exciting, atmospheric way. The place seemed to be sleeping and utterly devoid of people. I made my way down to the sea front and pulled the stop button on the tractor.

There wasn't a soul about there either, at least not that I could see. Only the part of the beach right in front of me was visible but the sea was lost and silent in the fog. I decided not to make my dinner here after all. The thought of standing alone in the thickest fog I'd ever seen, cooking beans in the back box of a tractor, on the edge of a beach, in a deserted village, just felt too weird. There was a tiny shop down near the front, which, amazingly, was open, so I turned around, parked in front of it and went in and bought a postcard of the village and beach. This meant that finally I was able to see what Llangrannog really looked like.

Leaving the village I roared steeply upwards – sparks-a-flying – out of the wet blanket of fog, arriving back in the middle of the sunny afternoon. I switched the lights off – congratulating myself on having thought to do so because sometimes my slow-wittedness amazes me – and went on my way, chugging along between the winding hedgerows again.

When nature calls on quiet lanes, peeing isn't a problem because you can just go in the grass verge. I suppose it's best to turn the engine off and then you can hear if a car is coming. It is mildly embarrassing being caught squatting with your pants around your ankles. The problem I have is that because of my knee injury, I can't squat properly so I have to hold one leg straight out to one side. This means that the pee comes out at an angle and it has taken years of practice in order to learn to avoid peeing into my shoe. Being a heavily pregnant, balaclava-clad tractor driver is one thing, but stinking of pee as well is just a step too far.

High up – about a mile from the sea – was a small farm and campsite. I drove into the yard and found the family who owned it. The chap turned out to be a semi-retired farmer. So he knew a thing or two about tractors and took an instant interest in mine. I kept trying to bring the conversation around to the subject of whether or not I would be able to camp here tonight, but it wasn't easy. Finally, he waved a hand in the direction of the camping field and told me that the loo was in one of the outbuild-ings across the lane. Then he refused point blank to take any money. I tried to get him to take the fiver I had in my hand because I didn't want him to think that I'd chatted to him for fifteen minutes in order to get a free night's camping. Perhaps he allowed me to camp without paying because I was friendly or because I was pregnant or because I was a fellow Welsh person or a fellow tractor owner, I shall never know. Anyhow, as I had missed my beans on toast at Llangrannog, due to feeling too much of a fool to sit in the mist cooking, I now felt really hungry and couldn't wait to cook. So without further ado I drove straight into the camping field.

This was a small square field, occupied by a big car and a top-of-the-range caravan, complete with an awning and one of those huge gas barbecues parked outside. I waved at the couple who were sat in the sun on their deckchairs alongside the gas barbecue and they waved back. I drove to the farthest point away from them, as was possible in a small field, and pulled the stop lever on the tractor. All fell silent and I realised that I was seriously tired.

I proceeded to get together the things I needed to set up my home out of the tractor, while wondering if the respectable-looking retired couple with the caravan were thinking that I was something to do with maintenance. Also I wondered if they were surprised to see a stove, a tent and a fold-up wooden chair come out of the carrier of the tractor. I wondered, as they watched me, if they thought I was pregnant or if they just thought I was fat. And if they thought I was pregnant, did they too think that I was running away? I had a sort of fantasy where they came over, made friends and asked me to join them at their barbecue but that didn't happen. People are just too reserved in Britain. But then, perhaps I wasn't forward enough. Maybe I should have walked right over to them and introduced myself.

Multi-tasking, I put the egg noodles, the tin of tomatoes and the tin of tuna all in a pan and placed them on the stove, while I put the tent up. In forty-eight hours I had become an expert at all this camping malarkey. I knew where everything was and what order to do it all in. In the time it took to boil the egg noodles, I was able to set up home, complete with my bedtime book *The Matter of Wales* resting ready on my pillow.

Just as I was about to eat, the smart but casual caravan man walked back from the shower, carrying a little zip-up bag and a towel neatly rolled up under his arm. I waved again and he altered his course and came over. 'On your holidays with a tractor, are you?' he said. 'Yes,' I said, suddenly aware of how stupid it sounded when put like that. I tried explaining the sheer wonder of being alone on a tractor holiday, what a great way it was to see the country, but he didn't look in the least bit convinced. I couldn't see him going back to his wife and saying, 'You know what, darling, I think we should flog this flashy caravan and get a couple of old tractors and a tent and take off like that.' As the man left to go back across the field to tell his wife what I was playing at, I was left thinking that he was probably thinking that I was some kind of a weirdo.

I'd brought a mobile phone with me, which I vowed not to use except for emergencies. I asked people not to phone me, unless it was a crisis moment, because I didn't see the point in getting away from it all by sitting on my tractor going, 'Yeah, yeah, I'm just sitting on the tractor now. . . .' down a phone to someone. If it hadn't been for the pregnancy issue, I wouldn't have taken the phone at all but it was something that Father-to-be had really wanted me to do and since it seemed so important to him, I agreed as long as I could leave it switched off. I had tried to turn it on once a day, just in case there was an important message, but so far there hadn't been a signal on the phone at all.

Initially, I was surprised that over two days I hadn't once found an area with a signal but then I realised that although I had travelled continuously for two days, I hadn't actually gone that far, in terms of miles. In two days, my world had obviously changed to suit the pace of the tractor and I felt like I had travelled huge distances. When you are following an OS map, it seems like you are moving far and fast but when you look at where you have been on a road map, it looks like you've barely moved.

Anyhow, there was a phone signal at the campsite near New Quay and I decided that, since it might be the last one I might see in a while, I'd better give Father-to-be a quick ring, just in case he was worrying. When I got through his first question was, 'Is the tractor going okay?' I was glad about his choice of words because it relieved me to know that he wasn't sat there fretting over me (although he was sitting there fretting over the tractor). Having someone worry over you unnecessarily is just patronising, I think, so I was glad that he saw no reason why a woman shouldn't drive a tractor, pregnant or not, if she wanted to. And knowing that he had every faith in my ability to handle myself made me like him all the more.

Before bed, while it was still light, I followed a path

that headed in the direction of the sea. I crossed a couple of fields and still seemed to be no closer to the cliffs when suddenly I developed a violent stitch and had to sit on a bank under a hedge until it passed. Then I decided that the sensible thing would be to return to the tent, in case it happened again. I had already asked the midwife what the stitch meant and she said that it meant I was probably rushing about too much. I would hardly describe walking a slow half-mile as rushing about, but there you go. Feeling regretful, because I still hadn't had a good look at the coast, I waddled my way back to the tent and took to my bed.

DAY THREE

Feeling like you don't fit in, back to the wilds,
Tregaron and a cute pair of bachelors . . .

E ASING myself uncomfortably out of the tent was
similar to being born, or so I thought as I crawled,
groaning and blinking, out into the sunny morning. I'd
slept like a log again, probably because of all the fresh
air, but all the same it would take a good half-hour for
my face and spine to straighten out. I went to investigate
the shower in the shed and, at the last minute, decided
not to have a shower after all. I'm not sure if this was
because I couldn't be bothered to cope with a tempera-
mental campsite shower or because showers are modern
conveniences, which meant that to use one while on a
rough-and-ready trip would be cheating. Either way, I
left still smelling of diesel because who was going to
smell me anyway?

Packing everything away each morning was already
becoming a tedious experience. I had become so effi-
cient that I had started to try to beat my previous time,
which was stupid since none of my holiday was meant

to be anxiety-fuelled in any way. The reason I found myself rushing to pack away was, I think, because I couldn't wait to get back on the tractor. Even when you are camping under the stars you still have to prepare food and wash up, which is, in a way, just like being at home, except with worse facilities and more limited ingredients. On the tractor, however, I was completely free of any such tedium and all I had to do was enjoy finding my way through the delightful maze that was rural Wales.

I wanted to see the sea and this want was further fuelled by the fact that I had failed to do so last night at the campsite and yesterday afternoon at Llangranog. So I took the back lanes to New Quay, following directions given to me by a small woman with the biggest breasts I have ever seen, and eventually I came to the sea front. It was deeply touristy and I felt amusingly out of place. I drove around, then parked up and sat on my tractor looking at the view, alongside a number of other people who were also looking at the view, except that they were sitting on park benches and not tractors. I suddenly felt that I needed to get out of New Quay, pretty as it was, because I just wasn't the same as all the other tourists. I was wearing far too many clothes for the time of the year and I was on a tractor. So I'd seen the sea and now I felt that I needed to get away from the public and back into the sanctuary of the middle of nowhere.

Back in the open countryside, where hedges and

fields stretch as far as the eye can see, I felt at home again. I felt like patting the tractor and saying 'good boy' for not having let me down yet, but I'm never really sure if my tractor is male or female, so I didn't. It doesn't even have a name, my tractor, it's just referred to as 'the thirty-five', which is a shame really because it's like calling your mate '1.7-metre-tall human', rather than 'Lucy' or whatever it is your mate is called. I've never found the right name, or even the right sex, for my tractor. I tried to refer to it as Megan in the beginning because that was the name of the lady from whom I bought it – but it just didn't seem enough like a Megan to make the name stick.

The tractor was an absolute star. It was just chugging along the lanes, hardly using any diesel. In fact, I hadn't even bothered to refill it yet, partly because it wasn't that low and partly because Father-to-be had filled the jerry-cans right to the top, which meant that they were quite heavy to handle. I thought that eventually I'd just ask a passer-by with a boiler suit to do this, unless it became absolutely necessary for me to do it first. I could imagine what the midwife's views on the lifting of jerry-cans full of diesel would be, so really, unnecessary weight lifting was best avoided if at all possible.

Aside from its wonderfully low fuel consumption, the tractor had run for hours and miles without missing a beat and had never failed to start when asked to do so. Because it had been so reliable, I had developed a new-found bond with it, as though it was my loyal steed, my dependable, easy-going companion.

If you're in the right area the single-track roads just go on for miles and miles. The hedges were billowing with the frothy white flowers of cow parsley and may blossom, which swept the road around the sides of fields and past the little farmyards with their round-topped barns and their sheepdogs sunbathing in the dusty yards, tongues a-lolling. Un-sheared sheep lay panting in shady gateways and in buttercup-riddled paddocks ponies swished their tails over their sunbathing, bob-tailed foals. It felt like this was the Wales of forty years ago, like I had gone back to a time before I was born; perhaps even back to when my tractor was new and ultra-modern.

In fact, the whole area seemed so peaceful and un-occupied that it felt as though I'd failed to spot a sign earlier down the road: one that had read 'Wales is Closed Today'. The only people I saw were farmers, walking across their yards and looking up, wondering who was going past on a 35. Farmers don't bat an eyelid at cars because they often contain people simply passing through. Tractors, on the other hand, never just pass through – tractors typically belong to neighbours so farmers recognise those tractors as being local. I passed countless farms where someone would stare in puzzle-ment at my tractor and me. I would wave and they would wave back in a slow and puzzled way, clearly thinking, 'I don't know who the hell that is but I'll wave anyway.'

Later in the afternoon, as the heat began to ease off a little, I could see into people's gardens from my great vantage point on top of the tractor. The busy gardeners of Britain were beavering away with their mowers, rakes and hedge-trimmers. It was hot and, though the pace of the tractor did cool me down a bit, I still felt that I needed a swig of water every few miles. I had brought lots of fruit to feed on as I went and there was nothing to stop me from pulling up on a verge and cooking a meal if I needed to but, as I was enjoying the chugging-past of the landscape so much, I had to force myself just to stop long enough to brew a cup of tea.

You might expect to get funny looks sat on a fold-up chair beside a tractor on a grass verge with a cup of tea in your hand but it was such a quiet area that not a single vehicle passed me as I sipped my drink. Two cups of tea and half a packet of biscuits later, I said, 'Right, let's get going then, shall we?' to the tractor, before realising that these were the first words I had uttered all day. There hadn't been anyone to talk to, which might explain why I was now addressing my tractor as a fellow human. It struck me that after several weeks of this a person could entirely lose the art of intelligent conversation.

A few miles later I found an opportunity to test my conversational skills when I saw what looked like the workshop of a rural agricultural mechanic. The doors were open and a man was fiddling about with the wheel of a van in the entrance. I chugged over, turned off, said 'Hiya,' and told the man that I was looking for Cilau Aeron. This was a small village that I didn't need to visit but knowing its whereabouts was a way of finding out if I was following the road I believed I was following on the OS map. The chap gave me directions for a mile or so, then said, 'When you see a man in front of you, up a ladder, painting the wall of a house, then you're nearly there. You can ask him the directions for the rest of the way.'

I went on my way tickled by his instructions. A small community where everyone knows what everyone else is up to is the only place you would ever be likely to hear directions like those. So feeling like Alice in Wonderland, or someone else on a strange mission who is being guided by peculiar signs, I chugged down the road towards the man painting the wall.

I spotted the house easily enough: it was facing me, part-painted and there was a long ladder leaning up the wall with a large tin of paint on the floor beside it. Clearly the man had gone indoors for a cup of tea – something the workshop-man hadn't bargained on – so I never got to ask him for the rest of the directions. This wasn't a problem because by now I had now worked out where I was in relation to the map. All the same, actually seeing the man up the wall would have brought a pleasant sort of closure to the little episode.

I decided that I would head in the direction of Tregaron. It was a small, remote market town with the Cambrian mountain range to the north-east of it. I would be able to buy some vegetables and maybe some cheese there, which would make a nice change from having only tins and noodles for tea. On my way towards Tregaron I saw that I would be crossing a part of the Sarn Helen, which is a Roman road reputed to travel the length of Wales. Sadly, the part of Sarn Helen that I was going to travel is now a B road, like any other B road, except for the fact that this one is really straight. The part that's up near my home in the north is a rough track, most of which is in much worse repair than when the Romans were about – there's no way you could get a car down it at all. Personally, I would rather have driven the Roman road that you can't get a car down but, on this occasion, I had no choice in the matter.

Tregaron has a busy history. *The Matter of Wales* informed me that it was once famous for wool and was home to a vast number of tradesmen and women who were part of that industry, such as tailors, hosiers, woollen-mill workers, cloth dealers, seamstresses, hatters and shawl makers. Tregaron was to wool what Sheffield was to steel. Suitably positioned in the middle of nowhere with nothing but sheep all around, Tregaron was an ideal location for this sort of business. Today, however, the advent of man-made fibres has meant that wool is virtually worthless to the sheep farmer.

Somebody somewhere must still be making a buck out of wool though because if you went to buy a wool carpet it would cost you an arm and a leg.

The Tregaron of today is an altogether quieter place than it once was but it is by no means a ghost town. It's a small place but it's got a huge rural catchment area, meaning it's still very much alive with people going about their everyday business. Yet I couldn't help but feel a twinge of regret for the fact that the wool industry had died a death because, after all, wool is a wonderful sustainable product. A wool coat lasts a lifetime – albeit a rather itchy lifetime – whereas a shell suit melts the first time you drop some fag ash on it. Perhaps that's the trouble, though – perhaps people don't want things to last a lifetime any more.

Arriving in Tregaron, I felt like I'd arrived in Piccadilly Circus; I'd been wandering the lanes for days and now this tiny little market town seemed so bustling. All these people and so much traffic was a shock to the system when you've only had the company of hedges for days at a time. I drove right up onto the square and parked outside the Talbot Hotel. As it happened there was a bench and a table outside, so I went in and bought a glass of Felinfoel ale and sat down outside to drink it, enjoying an excellent view of the square as well as my tractor, which was draped with all of my belongings.

Tregaron seemed built-up and busy compared to what I'd been used to and, yet, I didn't feel out of place being on a tractor. It appeared to be perfectly acceptable to ride into town on a tractor here, whereas in New Quay people had been pointing me out to their children, saying, 'Oh look a tractor', which had made me feel, somehow, like a fool. Here, though, I saw other tractors passing through – more modern ones, yes, but tractors nonetheless – and the sight of them made me feel quite at home.

Going camping with a tractor was a whole different thing to riding through a market town on one, however. Camping with a tractor separated me from the masses somehow, which was a pleasant but also rather odd feeling. The notion of tractor camping would seem strange, especially perhaps, to country folk, who never quite seem to see the point of camping in the first place. Town folk can't get enough of camping, even if it is with their cars safely parked next to them, whereas people like farmers, who are out in the fresh air every day, have no desire whatsoever to sleep the night in a field.

So, I concluded that it was just about okay to drive a tractor for pleasure – people who go on tractor club road runs do it all the time – and it was definitely acceptable to go camping, but to combine both pastimes wasn't normal somehow. It was as though I had broken the rules and dabbled with the set order of things provoking a slightly lonely sensation that comes from not quite fitting in. If you are a bricklayer who also happens to be a ballerina, then you will know what I mean.

I wondered if there were any circumstances where it would be considered reasonable to go camping with a tractor – such as if you had bought a tractor in South Wales and you couldn't arrange transport north for it, so you were driving it home and you took a tent with you so you could stop overnight? That might be a realistic story to tell anyone who chatted to me but there is a reason that I wasn't going to lie and tell someone that I'd just bought the tractor from a guy in Pendine. In my experiences, rural people know loads of other rural people and there is a good chance that this 'someone' happens to know every guy in Pendine with a tractor and he'll want to know your guy's name, and so on. So it's best not to lie, especially in the countryside because people want to know where exactly you're from, who your father is, where you went to school – and eventually you will get caught out.

I learnt about the hazards of telling lies back in primary school, when I pretended to another child that my family ran a stud farm, full of (literally) fabulous horses. After a few weeks, it became quite complicated and I dearly wished that I had never started the wretched fabrication. I couldn't keep up with all the questions I had to answer about all the different horses. I thought I might have to start a 'lie book', just so I could recall what I had said.

The next thing I knew, the girl started expressing an interest in getting her mum to bring her up to see all my wonderful make-believe horses. So stressed out was I by my lie – but having gone too far into it to now say it that it was only a joke – I began to dread going to school. I was caught out, and came clean but by then it was too late. The girl went right off me, of course, and she subsequently liked me far less than if I'd just been someone who didn't own a stud farm. Because I was worse than that, I was someone who not only didn't own a stud farm but was also a liar. So great was the shame that I don't think I've told a lie since, as long as you don't count exaggerations of course.

While I sat there with my beer pondering the many different reasons why people tell lies and studying my OS map at the same time, a lively looking man in tight denims with a fag packet sticking out of his top pocket came weaving his way through the parked cars towards me. He was waving his arms up above his head in mock anger and shouting something like 'cars, cars, bloody cars'. He headed towards me smiling broadly and leant over me saying, 'They're everywhere!' Then he threw his arms up again, laughed out loud and headed into the pub. I smiled and nodded, glad that he hadn't required a response.

When I went into the pub to take my glass back, I stopped at the loo and, for the first time since North Wales – however many days that had been – saw my face in a mirror. I was looking more like a bag lady than ever. The wind had given my hair a good shake and, thus, the full-on, mad-woman-in-the-attic look had been achieved, whether I liked it or not. Then there were the smudged oily marks on my face, which looked a bit like bruises. With that deranged look and the pregnancy, there was probably little chance I would be chatting to anyone long enough to need to give them an account of what I was up to. No wonder the passing nutter had come straight over to me outside the pub, he probably thought that he'd found a kindred spirit. I wiped off the grubby marks with a piece of loo roll and left.

I headed over the road to the Spar shop. Imagine, I thought, driving a tractor around the world and finding a shop like this in every town: how awful that would be. When I came out of the shop with my carrier bag full of stuff that I didn't really need, I spotted a council roadman peering into a hole in the ground at the rear of his lorry. Thinking that a local roadman would be an ideal person to ask about roads, I went over and asked him if he knew of any green lanes, or roads without tarmac, in the area. He didn't really know of any and he seemed bewildered by my quest, as if while he was slaving away to keep the roads in good nick, here was this woman stupidly searching for an ill-maintained dirt track to drive along. So I thanked him and went back to my tractor, feeling slightly peed off about being stuck with the B road that headed toward Devil's Bridge.

Tregaron would probably be a great little town in

which to have a few beers on a Friday night and there's no doubt you would meet some colourful characters but, alas, being pregnant in charge of a tractor I was forced to leave the bright lights behind me and get back to the fields and the hedges. While I was looking forward to being in the open countryside again, I wasn't keen on the prospect of driving on a B road, and I regretted the fact that I'd made the decision to come to Tregaron without first looking at where to go next on the map. Heading back a few miles wasn't an option for an impatient sort of person so I decided just to make the best of the B road after all.

Just as I had imagined, this was the sort of road frequented by boy racers. They tore by, passing me at sixty, weaving in and out of the traffic. Once I heard the screech of brakes, as four twats in a Nissan Micra almost ploughed into the back of me on a blind bend. 'This wasn't the sort of chilled out holiday I had in mind when I booked!' I heard myself saying to the holiday rep. I stopped for a breather in a lay-by and suddenly felt old and tired, sitting there tut-tutting about young people and the way they drive. Oh lord, save me from becoming a miserable old git, just yet.

With the traffic whizzing by, I sat on the tractor reading the leaflet that I'd picked up from the tourist information place in Tregaron. It was all about Cors Caron, the huge wetland reserve that I was now parked alongside. This area was once used for peat extraction and, in an effort to repair the damage caused by that, the powers-that-be have now designated the area a sanctuary for all the various sorts of wildlife that enjoy living in boggy places.

A man passed me taking his dog for a walk and, after saying hello and exchanging other pleasantries, he told me that he was going to walk along one of the footpaths that goes through the wetland. I recalled that Jan Morris's book had told me that this place, Cors Caron, was the only known home of the British black adder, so I asked the guy if he had ever seen one of them while he was on his walks here in the bog. He looked at me a bit strangely and I was forced to explain that I was referring to a type of snake, not to be confused with a comedy character played by Rowan Atkinson. He continued to look at me oddly, as if to say, 'What are you? Some kind of tractor-driving, snake enthusiast?' Then he said, 'Aye, it's full of bloody snakes. . . .' and wandered off along the footpath.

Further on, out in the open rolling hills, the B road seemed friendlier but driving at full pelt – or the equivalent for a tractor – was becoming loud and tiring and, despite my coat, I was beginning to feel the cold. I suppose I had just been sitting still for too long.

Once again, there were no campsites on the OS map, though I did come to a pub with a campsite out the back. I stopped but the pub was shut and there were a few lads hanging about outside, presumably waiting for

At Pendine Sands, ready for the journey.

Ladies relaxing in Tregaron.

Camping in Ffair Rhos on the third night.

The Edwards brothers, Twym and Cadwgan, were truly what
the Welsh call *halen y ddaear* – salt of the earth.

Crossing the Nant y Moch reservoir.

The top of the 'hill of almost certain death'.

Traversing the hills between Corris and Dolgellau.

At Whistling Sands, cold and windswept but triumphant.

it to open up. I suddenly felt that I didn't want to camp here anyway, it was all too close to the pub and the road and I had an unpleasant vision of five pissed-up lads trying to joyride my tractor in the middle of the night with me running after them in my nightie brandishing my tiny blunt penknife. . . . I think my pregnancy was making me feel a bit vulnerable, so with more than a twinge of regret that my nice cup of tea and my nice warm sleeping bag had been denied me, I soldiered on. Of course, I could have stopped in a lay-by to brew up but by now I was feeling so tired that I just wanted to stop for the night.

Further on up the road I came to Ystrad Fflur, which was once an abbey populated by Cistercian Monks but is now little more than a few ruined walls. I drove up to look at it but decided not to go in, as I was just too weary to be a fascinated tourist. Continuing down the road, I came along some sweeping bends and, far in the distance, across the open countryside, I could see that the road ahead of me went through a farmyard.

Even from a quarter of a mile away, I could see that a man was standing in front of the house, intently watching my tractor as it slowly approached him. I suspected that he knew about tractors and that he was probably thinking that no one around here has a 35, so who could be chugging along the road towards him? As I got nearer I could see that the chap who was watching me was wearing a jauntily placed beret and, from that and the way he was studying my tractor, I guessed that he had to be some kind of a vintage vehicle enthusiast. Partly out of curiosity, to see if I was correct in my assumption, and partly out of need for campsite information, I pulled up to talk to him and stopped the tractor.

Small in stature and hugely endearing, he had the loveliest way of speaking Welsh that I've ever heard. He was enthusiastic about my trip and still more enthusiastic about my tractor. He even called his brother over to come and see it. They made me feel like something of a celebrity and within minutes I was in their shed looking at the chassis of a 1940s lorry, which the one with the beret was in the process of restoring. I suddenly wished that I knew more about 1940s lorries.

The men I had just met were the Edwards brothers, a pair of bachelors whom everyone seemed to know, judging by the amount of cars that bibbed as they passed by. They were people that the Welsh call *halen y ddaear* – the salt of the earth – and they were a right pair of characters, too. When I told them that I was looking for a campsite, they immediately offered me their field, which was an amazing thing to have happen, as by now I was utterly exhausted and very, very hungry. I guessed that a pair of lads in a car with tinted windows and thumping techno music would never have had such luck at finding free accommodation. What was even better was that I felt totally safe to spend the night in this field because these two men were obviously gentlemen and while I know that psychopaths can disguise themselves as kind, respectable people that doesn't happen very often. You just have to go with your instincts on some things.

These brothers were extremely busy people; the one with the flat cap was the farmer and the one with the beret was the engineer. They were a fascinating pair. The chap with the beret was named Cadwgan and he was the chattiest: we talked at length about old vehicles, shows, farming and everything really. His way of speaking Welsh was a bit different to mine. Sometimes he would use phrases that I had to guess the meaning of in an accent that was mid-Wales with a hint of southern sing-song, but he was so well-spoken that it was easy enough to follow. I had quietly wondered what his thoughts were about unmarried pregnant women undertaking poorly planned tractor holidays. One might expect older people, especially those with old-fashioned lifestyles, to be sceptical about things like that; however, sometimes it is the individuals who are living an up-to-date life – those who are well-educated and well-travelled – who can be the most judgemental people of all. It was clear that Cadwgan thought that driving a vintage vehicle was a perfectly rational and, indeed, highly commendable thing to be doing with one's time, whatever the circumstances.

So, by chance, or by the grace of something, I had happened to choose to stop and talk to a kind gent – one who not only offered me a field for the night but also happened to be a vintage tractor enthusiast. What a find! I set up camp near a hedge and got the food in the pans, all the while reflecting on what a charmed life I seemed to be leading. It was only when I was sat down with a hot cup of tea in my hands and the tent was up and the tractor turned off for the night, that I could ponder over the day that had just been. I remembered how, before stopping here, morale had been at an all-time low because I had, in my tiredness, begun to wonder if I would ever find a safe place to put my head down for the night. That little low point had now been replaced with a big high. It's hard to explain how important the little triumphs are at a time like this but finding a place to stay before nightfall had made me feel brilliant again. Like a smooth, well-polished rolling stone.

Being free of all constraints is a wonderful thing: it's something that money can't buy. In fact, I think money actually makes it harder to obtain. Sometimes, the more cash you have the more complicated your life ends up being and, before you know it, you haven't got any time at all because you're too busy running around with, and for, your money. I remember how simple life was back in the early nineties when my home was a caravan that could be easily replaced for £200. I felt a bit like that again: like a simple happy soul.

I might have felt that everything I owned in the world was right here with me – my sleeping bag, my tent, my stove and my tractor – but that wasn't really the case. I had a (relatively) normal life waiting for me back home and this trip was just a holiday. But I thought that I would like to stay on this holiday, like someone playing cowboys and Indians, until I got a full taste of it; perhaps until normal life started to seem appealing again. I did wonder, though, what it would really be like to have only this in the world. To be a true rolling stone, roaming from place to place, not just for a holiday but for life. If I had been some ace agricultural-worker-come-mechanic, then I could have made a life for myself out of travelling around on a tractor. The tramp lifestyle has always appealed to me – not in a homeless alkie style, though lord knows at times even that can seem quite appealing – but more in the style of a skilled, travelling worker.

In my grandfather's time, there were people who were known as 'agricultural vagrants' who walked or hitched rides from place to place, working different harvests at different farms, helping with lambing at others and undertaking hedge-laying or general repairs in the wintertime. They slept in barns and would have seen more of Britain than any other manual workers could ever have hoped to at that time. I imagine a tractor-driving version of this: going from loch to hill, from barley fields to woodlands, seeing all of the countryside and becoming respected for your work.

You'd always have enough money for a beer, mates in

every region and you'd never have to worry about anything, except keeping your vehicle on the road which, if
you were a mechanic wouldn't be too difficult. It would
seem like the ideal life for a fit, young, gregarious and
perhaps eccentric agricultural mechanic, who, for one
reason or another, doesn't feel like putting down roots.
Sadly, I have a bad knee, am pregnant and am never
likely to be able to earn a living by mending people's
tractors, building barns or delivering calves. So the lifestyle of the wandering agricultural worker is unlikely
ever to be mine but I still think it would be a great
opportunity for somebody out there.

Just as I was serving up my dinner of vegetables,
noodles and cheese, Cadwgan came to the field gate and
called across, asking me if I wanted to come in to watch
Pobol y Cwm with them. This is a Welsh soap opera,
which happened to be a particular favourite of my late
father, and to find that the Edwards brothers were avid
viewers too made me smile. I had to decline though,
due to my dinner being ready on the plate, which was a
shame.

If I'm honest, I was quite curious to see the inside of
their bachelor home, expecting 1950s wallpaper and
lino, hand-me-down grandfather clocks and pairs of
white, curly haired dogs that sit on mantelpieces. I think
this was what I imagined the inside of the Edwards'
house to be like.

The place I had chosen to relax in the field was right

near the main road but behind a thick hedge. I could hear people driving by – they obviously knew the Edwards brothers and hooted as they passed, even if there was no one in sight. This seemed like a friendly thing to do. Luckily, because of the hedge, my little camp and I couldn't be seen from the road, otherwise I imagine there might have been a local inquiry into why a pregnant woman with a tractor was camping out at the Edwards' farm.

After studying the map, it looked as if my best bet for the next day was to go to Devil's Bridge and after that I would be able to get off the main road. Even though this was a really small main road, it still had white lines and rushing traffic along much of it, and I wanted to be back on my Postman-Pat-style, twee little lanes. Unfortunately, in order to get back on to one of those roads, I would have to drive over Devil's Bridge itself or go a very long way back.

Bugger, I thought. I suffer from vertigo, which means I really don't like crossing high bridges, especially while riding a horse or a tractor because you are even higher above everything then. My laid-back attitude, or rather my sloppiness, had caused me to not bother to get the next OS map out of the tractor box until my stop at Tregaron, by which time it was too late. I had wandered into a strange land where no single-track road ever heads north or even west or east, for that matter. There is only one road and it leads to Devil's Bridge. I tried to tell myself that the bridge couldn't be that scary and, in any case, I wasn't going back past Tregaron or further. No way. So tomorrow, I would just have to grit my teeth and go over the bridge.

Despite the hard bumpy field, and my crap carry mat, I slept well again, apart from the usual night-time pees. Two things I was really glad for that night were: a) that I wasn't visible from the road and b) that I didn't need a number two because one thing I hadn't thought to bring with me was a spade. On this holiday I just had to hope that my body would wait to have that particular need until it reached some convenient public toilets or a deserted forest even.

DAY FOUR

Vertigo, the grave of a great bard and Dolgellau

I WAS on my chair in the field, having just finished my breakfast of tea and two pieces of toast – one with peanut butter and one with Marmite, exactly like I have at home – when Cadwgan came over to see if I was alright. Conveniently, seeing that he was a man in a boiler suit, he kindly lifted the heavy water and diesel cans for me, and refilled my tractor. I experienced another wave of extreme wonder and gratitude at how well everything was turning out. We checked the oil and, once again, the tractor hadn't used a drop.

Before I left for the next leg of my journey, the other Mr Edwards, whose name was Twm, showed me around the farm and gave the new sheepdog, Juno, a quick practice herding the sheep in the field. Looking every inch the wise, Welsh shepherd, Mr Edwards stood there in a cool, slightly stooped pose with his flat cap and hands in his pockets, subtly directing the dog. It was a lovely image, and a lovely place, complete with a horse in the yard, which, I was told, was frequently used to get up to higher ground, almost like an environmentally friendly quad bike. We chatted about horses for a good while but I knew that I had to get going and do Devil's Bridge. So I said my goodbyes, which wasn't easy because, although I had only known these two people for a few hours, they'd treated me like an equal. They'd trusted me and they'd been my saviours and I felt quite emotional about all that. I felt like I could easily have had a tear in my eye saying goodbye but I kept a lid on it because I didn't want them to think I was bonkers.

As I drove off down the road, I rejoiced to myself about the fact that the Welsh language was still going strong and that human beings could still be relied upon to be kind and generous to one another. I marvelled over the charmed and wonderful life I seemed to be leading. I waved at people in gardens and nodded my head at several passing farmers in Land Rovers, whose occupants seemed mostly to nod at me first. I passed my time by singing out loud: songs about steam trains, mostly of the sort that Johnny Cash used to sing.

When I came to Devil's Bridge I promptly stopped singing. The place was shady and foreboding. Its only saving grace was that it had public toilets nearby because if there's anything that will bring on an urgent need for a public convenience, it's the thought of crossing a high bridge on a tractor. There was also a small shop just before the bridge and I glanced quickly at the bridge before hurrying inside. Stupidly making a last-ditch attempt at getting out of driving over the bridge, I found myself asking the lady behind the counter if there was another way across the river. She looked at me steadily and then said in a clipped, high-pitched tone, 'There is only this bridge.' I nodded and left, wondering why I had ever asked her that question when I knew full well from the map that there wasn't a better bridge down the road. What had I expected? That locals might know of a secret alternative to Devil's Bridge, one so secret that it isn't even on the OS map? And that this bridge might be a really low, safe bridge with solid edges that was only available to legitimate sufferers of vertigo, who also happened to be pregnant and driving a tractor?

Feeling like someone about to do something really brave I lined the tractor up with the bridge, hoping that the woman from the shop wasn't looking. I had decided to drive slowly over in a low gear, as near to the middle of the road as was possible without causing an accident. I flagged on the cars that were behind me, hoping that I might get the bridge to myself; the last thing a vertigo sufferer needs is to feel crowded at a time like this. I took a deep breath, gripped the steering wheel and chugged off.

I soon realised I should have chosen a higher gear because now I was on the bridge it was all taking far too long. I didn't feel like I could change up a gear since that would have meant letting go of the steering wheel with one hand and I felt too shaky to entertain doing that. So I just revved my way over it, very slowly, while gritting my teeth. Soon there were cars behind me, the occupants of which must have been wondering what the hell I was doing crawling along on the white line on a tractor, hunched over the steering wheel. I felt that at any moment the tractor might shed a wheel and plummet sideways, through the inadequate looking railings to the black rocky depths below, and me with it.

Before long – because the bridge was only short, no matter how slowly I drove over it – the whole ghastly experience was over and I was able to pull over to let the cars pass and breathe once again. On the one hand, I felt that I'd been extremely brave, while, on the other hand, I felt like a plonker for being scared of driving over what was, after all, a perfectly safe little bridge. What's so crippling about having a phobia is the knowledge that, to an outsider or a non-sufferer, you just appear bewilderingly stupid.

As I had hoped, I soon came to a lonely little lane that headed up into the hills towards the Nant y Moch reservoir. All was pleasant once again and as far as I could see on the map it looked as if I could go a really long way on this small lane. It even looked as though, later on, I might be able to get off the tarmac altogether and do some off-roading on a track. I was in positive spirits as I chugged up the road. I took my phone out of my pocket to check the time and saw, once again, that I had no signal. I hadn't needed to phone anyone, so the lack of a signal hadn't been a problem. However, it was interesting to note that just when I left all forms of civilisation and went into areas where I might actually need a mobile phone – should the tractor conk out, or my contractions start happening – then the signal would duly fade away, bar by bar.

As it was I felt fine and the tractor was still running like a watch, so I was able to enjoy the big open landscape with lonely farmhouses clinging onto the bleached grass and the shiny, wet-looking stones of the hillsides. It was colder up at this height and the light had become harder and greyer but it was still a very beautiful place. Before long, I came to the reservoir or, rather, the dam at the end of the reservoir. I saw with dismay that I would have to drive over the weird little road that was built into the dam and once again there wasn't a nicer way across it. I decide to pull up for a rest anyway. I took a couple of apples out of the saddlebags and wandered about. On the hill above the road there was a monument to our Welsh hero, Owain Glyndwr. It was there to commemorate Glyndwr's victory at Hyddgen in 1401 and it said, '*I gofio'r sawl a ddisgynnodd yn y frwydr*,' which means, 'To remember those who fell in the battle.' I felt a little shiver down my spine, thinking of warriors up here in these desolate hills battling it out with arrows and axes.

I always feel moved when I think of Owain Glyndwr,

TO COMMEMORATE
OWAIN GLYNDWR'S
VICTORY AT HYDDGEN
IN 1401
I GOFIO'R SAWL A DDISGYNNODD
YN Y FRWYDR

not just because he's Welsh but because he was on the side of the downtrodden. Nothing makes a better hero than one who has the courage to stand up against the powerful oppressors of his country and then beat them, against all the odds. Imagine a re-write of a typical old cowboy film where this time the Indians win and they kick the white man's backside and send him back home. Well, at this spot in 1401, the Indians did win for once. All right, Glyndwr didn't go on winning forever but, even so, it is said that his grave – the location of which, incidentally, has never been found – is in the heart of every Welsh person.

So sobering was this little slate plaque, with the dramatic barren hills all around me, that when I got back to the tractor I just climbed on board and drove straight across the dam without any more fannying about. I even managed to stop part-way along and take a photograph of the tractor parked up on the dam. It wasn't half as bad as I imagined and once I was half-way across I was able to admire the huge expanse of the reservoir, which stretched for miles across what was otherwise just moorland. The huge, dark form of the water branched off here and there, winding around outcrops, as though the mountains had been filled almost to the brim with water. What a weird place. It was utterly deserted, bright and sunny, but also extremely cold. I chugged along the lane that wound around the edge of the reservoir, marvelling at the place and feeling at the same time that it was strangely eerie.

I came to an abandoned looking cottage, not far from the water's edge and it struck me that, before the waters rose, there must have been more homes here. To think of a lost civilisation that might be down under the water was both a fascinating and a morbid thought. If a ghost town is creepy, then an underwater ghost town is creepier still. The odd thing was that, even above the water, this place was completely deserted. Single-track roads usually have some people living along them and you would expect to meet a car from time to time. But I hadn't seen a soul up here, a fact which had enhanced my experience of this landscape, making it seem not so much peaceful but achingly lonely.

When the lane began to take me steeply downhill I felt the air warm up and I saw that there were masses of bluebells in the trees, peppering the hillsides above and below the lane. The sight of vivid, dizzy-blue acres of bluebells is one of those things, like a perfect sunset, that never fails to lift the soul and to impress. As the blue haze was above me, on the steep hillside, I felt lucky that I had a perfect view of the display. After the haunting coldness of the reservoir, this place seemed warm and colourful.

I soon found the entrance to the tarmac-less track that I had seen on the map. It started off as a road leading to a farm, which was nestled at the foot of a steep hillside. I decided to call there to ask someone if it was okay to use the track because, in my experience, it's always best to keep on the right side of landowners. In the yard was a large, modern tractor with weights on the front. In comparison, my tractor looked like a Dinky Toy. I knocked on the open door of the farmhouse and made the decision to speak in Welsh, because, judging by the set-up, it was almost definitely the home of a Welsh person.

A no-nonsense farmer's wife eventually appeared and she said that I could go that way and that it was definitely passable with a tractor because her husband went that way to get to the top of their land. She asked me where I was trying to get to and I told her that, eventually, I was going to the Llyn Peninsula, and she just said 'Oh'.

I bade her farewell and was walking back towards my tractor when she called after me in Welsh, '*Be yn union 'di pwrpas y siwrne 'ma?*' – 'What exactly is the point of this journey?' Which turned out to be a question that I was going to ask myself some ten minutes later.

I started up the narrow, stony track and I knew it was going to be steep because it had to go over a hill and down the other side. As I progressed, however, it became steeper than I had originally bargained on. I had thought it might weave its way up the hillside but it didn't, it went almost straight up. Soon the front wheels were beginning to bounce up off the track and I found myself crouching right forward over the steering wheel in some hopeless attempt to weigh down the front of the tractor. Not only was the track on an extreme gradient but, also, the land to the bottom side was dropping away in an alarming manner, plummeting down to the narrow valley below, acres away.

When the front wheels come up on a tractor, it means that the steering wheel no longer works and the tractor can rear up and lurch off sideways. This isn't nice at the best of times but is an even worse prospect when you are near the top of what might as well be a precipice.

The only way to steer when a tractor is lifting at the front is to uncouple the brakes, so that they can work independently. Then you can steer in a rudimentary fashion by applying one brake at a time, because by momentarily stopping one back wheel you can make the tractor lurch to that side. In order to do this I would have had to stop and bend down to uncouple the brake pedals and I felt that I dare not stop, as the tractor was only just making it up the slope. If I were to stop, then the tractor might not start again and, anyway, it might not hold on a slope like this.

Since reversing back down would have been even more dangerous than continuing, I decided it was just one of those moments where you have to swear lots and ride it out. I could have kicked myself for not thinking to uncouple the brakes in the very beginning. I now knew, in no uncertain terms, what those close-together squiggly lines on the OS map had meant. All I could do was to lean forward as much as possible, pray that the tractor would continue its steady ascent and that the gradient wouldn't get any worse; whilst all the time spinning and lurching upwards, swearing under my breath, 'shit, shit, shit', and thinking, what the hell am I doing here?

Finally, it was over. I wasn't quite at the top of the mountain but at least I'd climbed the steepest part and was able to stop. I paused, breathing a sigh of relief. Then I took my phone out of my pocket and once again there was no signal. Not that I had need of a phone anyway, unless my water broke, in which case I would have preferred to take my chances walking down to the farm, rather than face being rescued by a helicopter, which, for a vertigo sufferer, would just about be the icing on the cake. I realised my hands were shaking a bit and that I was craving a cigarette – yet another one of the pleasures denied to those who are pregnant. Though, really, compared to driving up a mountainside on a vintage tractor, smoking a cigarette seemed a pretty innocuous pastime. Just imagine what the midwife would have said if she'd seen me coming up this track!

After a few minutes and few more deep breaths, I began to feel a lot better and was already starting to look upon the episode as an amusing adventure instead of the terrifying experience it had been. Ah, the wonders of fast-acting Rose-Tinted Spectacles! Deep down, though, I knew I'd been a bit stupid. In my quest to avoid the main roads and their associated perils, I had unwittingly put myself into a potentially lethal situation.

I wondered why the farmer's wife had said that it was alright to drive a tractor up here but then I realised that I should have observed that the tractor her husband drove was a hulking four-wheel-drive machine, complete with weights to hold down the front. My tractor was a dinky vintage number, complete with a pile of weight on the back. Anyway, it was done now and I could relax and enjoy the views from this most excellent vantage point, which I had struggled so ridiculously to reach.

A moment later, I continued on my way and came face-to-face with a gate which led onto fields which sloped, gently, down the other side of the hill. It didn't look like the track went through the gate, as I might have expected. Instead, it veered off quickly to the left just in front of the gate and followed the fence along the top of the hill. Tractor tyre tracks could clearly be seen leading this way, so I followed them. I was heading along the top of some very steep land and, while it was alright in the beginning, the track soon got too close for comfort to the steep edge and I began to feel my heart racing again. The ruts I was driving in became deeper and deeper. The farmer had obviously been up here in some really wet weather. Unfortunately, the tractor that had made the ruts was wider than my tractor, so I could only have one wheel in a rut while the other wheel had to drive awkwardly along the lump in the middle. Every so often one of the struggling front wheels would slip off the lump and land, with a crash, back down into the rut. As I was on top of what felt like the steepest field in Wales, the crash really made me jump each time it happened.

I couldn't even drive in the field alongside the tyre tracks because the ruts were on the only bit of the field that wasn't located on a 45-degree angle. I started swearing again and began to think that perhaps the track didn't go this way after all and I was just following in the wheel tracks of some farmer who loves the dizzy heights. But there was no way I was giving up the relative safety of the ill-fitting ruts to turn around in the steepest field in Wales, so I had to carry on. Finally, I came to a flat piece where the track came to an abrupt end right beside a big-bale feeder. I only had one choice: I had to bump and slide my way all the way back along those too-big wheel ruts, across the top of the steepest field in the world, and back to the gate. Not only was it a pain but I expected that any minute I might hear an angry voice telling me to keep to the track and not to go driving around on private land.

At this point, I felt like having a bit of a weep but I pulled myself together and got back to the gate that I should have gone through in the first place and said 'phew', several times. The track on the other side of the gate was barely visible and to start with I couldn't be sure if this was the way marked on the map or not, but at least it was a gentle slope this time. Soon I came over the brow and an amazing view of the Dyfi estuary opened out in front of me. I parked up and took a few photographs, experiencing the kind of elation that one often feels when one has had a close encounter with the jaws of death and lives to tell the tale.

So I had a blissful drive down through the fields on, this, the other side of 'the hill of almost certain death'. Speaking of death, the OS map showed that, not far from where I would join the tarmac road, was the grave of Taliesyn, a sixth-century bard, one of the first to be published in the Welsh language. He wrote deep and lengthy pieces, which even Welsh people seem to find a tad complicated to read. Taliesyn's birthplace just so happens to be close to where I live, so it was interesting to discover that his grave was around here and I was curious to see it.

The track soon led me to a gate and from there onto a narrow tarmac lane with grass up the middle. Because so far the track hadn't behaved like it promised to on the map, I wasn't entirely sure where I was. The OS map showed another track starting half a mile away, which I had planned to find, but I couldn't be certain of my actual whereabouts, so I thought I'd better ask someone. I wasn't sure I wanted to do any more off-roading because the last bit had been rather extreme but it would still have been nice to find Taliesyn's grave.

I could see a farm down below me so I chugged over and into the farmyard. Apart from the obligatory shifty-looking sheepdog, which was expressing more than a passing interest in biting my tyres, the place seemed deserted. I dismounted and walked over to knock on the door of the house, all the time keeping an eye on the dog, which was, now that the tyres had stopped moving, forced to look elsewhere for something to bite. It was skulking around behind me, looking intently at my feet, waiting for an opportunity to bite my heels.

I got to the door by walking sideways and not breaking eye contact with the dog because I had discovered that he only moved if you took your eyes off him. He stood glaring at me in a ready-to-pounce position, while I knocked and knocked again. I swear I could see him thinking, 'You can knock all you like, but there's no one coming to save you. Ha ha!' He snarled as I went to leave the doorstep and, because I thought I was quite likely to get bitten, I looked around for a stick but there wasn't anything about. I just had to move very slowly across the yard, one careful step at a time, talking to the snarling beastie, who obviously thought that I was a thief who shouldn't be allowed to leave the crime scene. Finally I reached the tractor and as I fired it up, the dog lost interest in me and started to attack my tyres again.

Although the dog and I hadn't exactly hit it off, I still didn't want to run him over, so I tried shouting *cer o 'ma!* (go away!) at him in an attempt to shoo him away from the front of the tractor. Just as this was going on a man swept into the yard on a quad bike and immediately the dog stopped biting my tyres and went over to his owner, wagging his tail. I hoped I hadn't annoyed the man by shouting at his dog. I explained that I'd been knocking on the door and, grinning, he said that he knew because he had been watching me from the top of the hill.

'Oh,' I said, glad that I hadn't managed to find a stick to throw at the dog after all and feeling ever so slightly foolish because he had watched, in amusement, my encounter with the snarling beastie. As I asked him where Taliesyn's grave was, he pointed to a hump on the grass verge just nearby. I said in Welsh, '*y boncyn?*' What I mean to say was *boncan*, which means hump, but

I didn't, I said something which sounded just like bonking, which is another kind of hump altogether.

I reddened foolishly. The man instantly recognised my mistake and burst out laughing. I really felt like quite the plonker, what with my inappropriate comments and the dog incident. The man was clearly finding me quite hilarious but I wasn't sure if he was laughing with me, or at me; still, as long as he was laughing, I supposed that it didn't really matter. He asked me where I was going and basically pissed himself laughing at my answers. When I told him which way I had just come and about the vertigo, he said, 'You went *that* way!' and howled some more. To try to make the whole thing sound more sensible, I said that I would probably write about the trip and ultimately sell the piece of writing. At this he laughed even more, managing to say between guffaws, 'A book about a tractor holiday, eh? Ha ha ha, bloody classic!'

Then, recovering himself – and perhaps feeling that he'd better stop taking the piss out of me – he asked me what I was doing about diesel and if I needed any. I decided it would make sense to buy some fuel from a farmer; after all, I might not pass a garage so it would save me having to make a detour to find one. He filled up the tank, which didn't take all that much doing because it hadn't used a lot since Cadwgan had filled it up with my diesel earlier (how long ago that now seemed, and yet it was only this morning!) Then he blatantly refused to let me pay for it. 'You might do the same for me one day,' he said in Welsh, and I said, 'Why, do you think you might do a tractor trip one day?' To which he replied, 'Not unless I completely lose my mind.'

Before I left, he told me that some scientists had once come to examine Taliesyn's grave to ascertain if it really is an ancient monument and not just a lump in the ground with two rocks sticking out of it that some joker, centuries back, had once claimed was the great bard's final resting place. You never know with country folk, sometimes they just make stuff up either because they are bored or because conning gullible tourists can be an amusing pastime. You always hear about how wonderful tourism is for Wales but, for many people, the tourists are the one big downside of living in a beautiful spot. Tourists make the idyllic places far too busy, they take all the parking spaces, they clog up the roads and they can't reverse to save their lives. No wonder people get pleasure from winding them up.

Like the old man who lived near my home who used to buy lots of those little plastic Jif lemons – the ones that come in yellow, lemon-shaped containers – which he then used to wire onto an old hawthorn tree in his garden. Walkers who passed by on the footpath outside his property would often ask him how come there were so many lemons growing on his tree up in cold, wet Wales. He would then inform them that this was a

Welsh Lemon Tree that was specially bred to produce lemons in cold, wet climates. Some people probably believed him and maybe went on to tell the folks back home about this rare and amazing tree. Those who were more clued-up and country wise would know that he was lying but they probably had an otherwise dull day coloured by their encounter with this mad bloke.

Anyhow, with regards to the Taliesyn grave, it soon transpired that these specialists had tested the soil in the grave and it was found to be earth from a completely different area, which meant that for one reason or another, great trouble had been taken to fill the grave in with some special soil. Which I guess could indicate that this really was Taliesyn's grave; either that, or it had been the work of some seriously thorough practical joker.

It turned out that I had passed right by the grave on my way to the farm because there wasn't a sign saying what it was and it did just look like a lump in the ground. I stopped to pay my respects to the bard and then left. The farmer had also been able to tell me where exactly I could find my next piece of off-roading and he had assured me that it was in no way as vertigo-inducing as the last one had been.

Since the guy had been something of a comedian, I just had to hope that he was telling the truth. As it happened, he was. The road was a lovely little track with grass up the middle, lined with wobbly stone walls and, most importantly, it was all on the level, meandering gently between meadows and oak forests.

Eventually, I came to a gate where I stopped and sat on my tractor, studying the map. A walker with a back-pack came towards me, said hello and asked me for directions. I held up the map and explained that I wasn't from these parts, which obviously puzzled the man, who, no doubt, hadn't expected to find a tourist on a tractor. He then asked me how I happened to be here on a tractor and I felt obliged to tell him, though I had begun to seriously dread telling anyone what I was up to since each time I had done so, my story had been met with a mixture of scorn and laughter. This man's only comment was, 'It must be very slow', and I wasn't quick enough to think to say, 'So is walking mate, but it doesn't stop you enjoying it!' I was able to offer him my map to look at but that didn't seem to help him. He persisted in asking me about the path ahead, as though he thought I was making the whole tractor holiday thing up, and I must be a local because I was, after all, driving a tractor. I couldn't offer any advice to this lost rambler because my 'vertigo-mountain' experience had started to make me feel that perhaps I wasn't very good at map reading after all.

Moving on, I soon found myself driving along a forestry track, surrounded by tall, dark, looming coni-fers. After winding around for what seemed like miles, I began to suspect that I wasn't on the right road any more because the road on the map had appeared quite straight. To make myself feel better, I told myself that perhaps the OS map people didn't bother to illustrate every single bend and, in any case, I had been travelling for ages and didn't want to start backtracking now. Later, when the forestry road ended abruptly onto a trunk road, I knew that I definitely hadn't been following my planned route. Bugger, I thought. Damn me for not turning back when I first suspected something was amiss and damn me for not being able to follow a map properly in the first place!

Well, now I was going to pay for it: this was a fast road with lots of sweeping corners, just what you don't want to be on when you have a top speed of something like twelve miles per hour. Going back would waste at least an hour and it had already been such a long day – what with the two lots of vertigo and all. What's more, the nearest campsite on the map was miles ahead, beyond Machynlleth, so I decided to brave the main road for a while because it was the only thing in the vicinity that appeared to head in the direction of Machynlleth.

As I was driving along I began to need a pee again. Tractor vibrations seem to enhance the pregnant need to pee all the time. Main roads are no good when you want to pee frequently because the passing traffic can quite easily think you are mooning them, which is quite embarrassing. Oh to be a man and be able to stand up and pee discreetly into the hedge without revealing your buttocks to the world!

Before long, I came to a place called Dyfi Furnace and I pulled onto a garage forecourt and bought a can of pop. I asked the man behind the counter about campsites in the area but as far as he knew there weren't any. This was a real shame because I would have liked to have stopped in the village to examine the huge waterwheel that was there and the lovely sounding Artists Valley, which lay above the village.

However, I felt that I couldn't waste time here because it was getting close to tea-time and I was anxious to secure a place for the night. Perhaps if I had taken the little road that led to Artists Valley, I might have met a kind landowner who may have allowed me to spend a night in their field but, this time, I wasn't prepared to risk it, with the knowledge that if it didn't work out, then I would be driving up the main road to Machynlleth in the dark. Funny how sometimes you feel gregarious and sometimes you really don't.

As main roads go, this particular one was all right. The height of the tractor meant that I could see right over the wall to the estuary below, which is something you can't do properly in a car. I imagined that a little boating holiday, following miles of rivers, could be just as wonderful as a tractor holiday. Around every corner would be a new scene and you could camp the night at beautiful, otherwise inaccessible, little shores and coves. In my mind, I created a holiday that was somewhere between *Swallows and Amazons* and *Deliverance*, containing only the best parts of each. I would be more sensible than the *Deliverance* lot – I'd get out and carry the canoe around any dodgy looking rapids and I'd certainly manage to talk my way out of being buggered by two deranged hillbillies.

The map informed me that further up the road I had the option of a back road to Machynlleth and, naturally, I was really keen to find this particular road. When I came to the junction I was looking for, sat there in a field, right beside the road was a tractor exactly like mine. So well positioned was the tractor that it could have been a sign for me, telling me that I needed to turn here. As I turned down the road I came to a man sat in a van in the gateway, looking at the parked 35 the other side of the gate. He seemed unsurprised to see another 35 appear in his rear-view mirror.

I waved at him in his mirror and he slowly got out of his van with a sigh, as if his life was nothing more than a constant stream of women on tractors, appearing in his rear-view mirror, pestering him for directions. He confirmed that yes this indeed was the back road to Machynlleth but that the main road was much more direct. I tried to alert him to the amusing coincidence that we both had the same tractors, but he pointed out to me that his tractor was in fact a 35X, and not a regular 35, which put me in my place because the 35X is generally thought to be the deluxe version. Anyway, he was polite enough, so I thanked him, bade him farewell and went on my way.

My heart sang again, I was back in Postman Pat territory: the tiny road went up and down and deep between the twisty, fern-clad old oaks, over little stone bridges, past wobbly moss-coated walls and gushing streams, and in between old, ivy-covered gateposts. I met a Land Rover with what looked like two farmers inside and had to reverse to a passing place because they had a trailer behind them. It should be written in the Countryside Code that all tourists should give way to locals, especially if the locals in question are towing a trailer. They smiled and waved a great deal, whether this was because I had backed up for them, or because they found the sight of me somehow amusing, I really don't know.

Later on someone who looked like a tourist reversed his car to a passing place for me and when I went to pass, he beckoned me to stop and asked me if I knew of a place where he could fish. I had to explain, once again, that I wasn't a local and I received that same surprised look, which was something I was now beginning to expect. I gave him my map to look at but that didn't help him at all.

Other than that brief encounter, the road was quiet and I had the world to myself – what a twee and tranquil world it was. I had a sudden craving for hot, stodgy food, probably because I hadn't really eaten since the Edwards brothers' place early that morning. I decided that once I arrived in Machynlleth I would go to a chip shop and treat myself. The day was drawing to a close and by now I was getting a bit sick of cooking my dehydrated food and opening my tins every night.

I arrived at Machynlleth to find a pleasant town with useful shops, lots of cafés and some charming architecture, including a graceful town clock next to which I was able to park. I think hunger and exhaustion were making me feel cold so I left my long, old, fake-fur coat on and walked up the street carrying my saddlebags. The saddlebags were quite heavy because they held everything important – money, maps, one bottle of water, my camera and a notebook. By the time I got up the street near the fish and chip shop, I was puffing and panting a bit and I felt flushed-hot and wobbly. I couldn't see anywhere to sit down so I put the bags down, took off my coat and leaned against a wall for a couple of minutes.

It was probably my pregnancy that was giving me a bit of a dizzy spell, coupled with the fact that I was rushing about carrying a lot of stuff and wearing too many clothes. Though the shops had closed, there were still plenty of people about and I felt that I'd rather not have a funny turn in public, so I continued up the street, past the chip shop and towards the public toilets. Thankfully, they were open and after a while in there I felt much better. I had hoped to make use of the mirror and straighten myself up a bit but, as it turned out, the mirror was one of those aluminium squares – the sort that's about as much use as looking into a piece of kitchen foil. You can check that you are still there – in the form of a fuzzy silhouette – but it's impossible to see any detail whatsoever. So I gave up on that and returned to the street to join the queue at the chip shop.

The man behind the counter was one of those jovial sorts, the kind who has mastered the art of witty and amusing banter and who can entertain the queuing customers so well that they don't notice how long they are having to wait. He was able to confirm that, yes, the nearest campsite was a couple of miles up the road, and that yes, I would have to go on the main road to get there.

I put the fish and chips into the saddlebag and headed back towards the tractor. I saw my reflection in a shop window and was sad to see that I looked more fat than pregnant and also that, due to the hat I had been wearing, my hair looked like the sort of style sported by medieval monks. In fact, with that hairstyle, my long black coat that wouldn't close and my huge belly sticking out, I bore more than just a passing resemblance to Friar Tuck – which is not a look that any female really wants for herself.

I had a small crisis of confidence and decided not to sit on the tractor eating my dinner while watching the street, which was what I had initially thought of doing.

Suddenly, I felt like I wanted to enjoy pigging out alone in the sanctuary of the middle of nowhere. So I left the fish and chips in the saddlebag, took the wheel, headed out of the town and roared up the road towards the campsite.

Shortly before reaching the campsite, I saw a place where I could pull off the road and I sat there, overlooking a river, still on the tractor, literally stuffing food into my mouth. I felt like someone who hadn't eaten for days and I was glad no one was watching because I ate like a glutton. Perhaps, I thought, the baby was having a growth spurt and my body was suddenly telling me I needed to eat. Even at the best of times, after eating fish and chips you can be left feeling bloated but when you're pregnant, it feels as though you're going to burst. Brimming with chips and purring like a cat, I rolled into the campsite and paid my dues.

Bending over to put the tent up I puffed and panted. I was beginning to feel extremely pregnant. I don't know why the midwife had tut-tutted about bumpy old tractors and rough roads when it was the bloody camping that was doing me in, especially all that crawling about on my knees. Setting up camp each night was beginning to get a bit tedious, probably because it always happened at a time of the day when I was hungry and tired. It was becoming like a ritual that I had to endure before I could be allowed to eat and lie down. Yet, once I'd actually pitched the tent, cooked my food, eaten my fill and was sat on the fold-up chair with a cup of tea in my hands, I'd feel as cosy and relaxed as someone at home in front of the telly. So with my soothing cup of tea, I sat beside my tractor and surveyed the campsite.

As campsites go, this was a really nice one: not too formal with plenty of trees and space and a river on one side of the field. There were a few other tents, a camper van and a couple of caravans, but they were all reasonably far away, and dotted about, as if everyone just wanted their own space. There were no humans in sight. I made my dinner, ate it and then I went for a wander down by the river. For the first time on this trip I felt a little bit lonely. I had seen some lovely sights and encountered some wonderful and amusing people but there was no one with whom I could talk about it. Most of the time I had been so happy to be the captain of my own ship, that it hadn't entered my head to want someone else there with me and that still held true, it was just that I had a curious lost feeling come over me.

Then I really began to notice the midges. I should have guessed that a woody glade near a river in June would be midge hell come evening time. I sat on a fallen tree, scratching and watching a couple walk across the field to the camper van. I noticed that a bloke with a bike had arrived at one of the small tents and was removing his panniers from the bike. I wondered who else was staying on the field tonight. This place was the most public place I'd slept in so far. In the other places I had either been alone or with the caravan couple near New Quay and they had given me a pretty wide berth. I must be an anti-social soul, I realised, because I tended to feel more comfortable alone than with strangers around me. I felt oddly vulnerable in a tent in a public place; after all, any wandering weirdo could pitch up a tent at a campsite and spot the lone female retiring to her tent at nightfall. . . . Paranoid thoughts.

Why did I suddenly feel like I couldn't trust the human race, especially when everything had gone so well for me? I shuddered and told myself that I was just tired and that it probably looked as though I was the weirdo on this campsite: sitting, as I was, alone on a fallen trunk in the margins, scratching and suspiciously surveying the other campers.

I got a grip and went back to the tractor to check it over for oil, water and fuel, noticing that there was a small pipe on the side of the engine that was leaking fuel slightly. Since it was nothing much, I decided that I would ignore it for now. The pipe looked quite delicate and amateur attempts to tighten it up with a spanner might prove disastrous. Instead, I put the kettle on, then puffed and panted as I unfolded the maps and laid them out onto the grass before me.

It wasn't long before I realised that I was going to have to retire to the tiny interior of my tent with this cup of tea because the midges were beginning to get the better of me. I think the pregnancy was making my blood even tastier than usual so I wore the fireproof balaclava that Father-to-be had made me bring, while I waited and waited for the painfully slow kettle to boil.

As I limped around in fast circles, balaclava-clad, trying in vain to get away from the biting cloud of horribleness, the man from the camper van came over and, in a polite English accent, said, 'The midges are awful, aren't they?' He handed me a small tin and said, 'Here, have this, we've got plenty more,' and told me it was an insect-repelling candle. I thanked him kindly and he trotted back towards the safety of the camper van, swotting at the air as he went. Thank the Lord, he didn't even ask about the tractor because I couldn't have been bothered to go into all that again now.

I got into the tent and laid myself carefully down, zipping the door on me, as far as it would go, so that just my head and one arm were sticking out of the tent and, in the acrid lemon fumes of the candle, I studied the map and supped my tea. I wondered if, along with cat poo, peanut butter, beer, heavy lifting and soft cheese, the inhaling of insect-repelling candle fumes was also to be avoided in pregnancy. Probably.

Then I discovered that this campsite didn't appear on any map I had. I was in unmarked territory somewhere out on the white margin that goes around the edge of the map. Today's map had ended with Machynlleth, whereas the next map up started just south of Dolgellau. Unfortunately, I had bought two maps of quite different makes and they didn't meet or, rather, there was a gap of about fifteen miles between the two. Being in between maps meant that I'd have to rely on asking for directions come morning. I could just hear myself saying, 'Excuse me, could you help me, I'm in between maps you see. . . .'

But whatever happened, I decided that I would continue up this lane, rather than head back to the main road. It would be my next mission, to get to Dolgellau the scenic way, no matter what. With that settled, I slept like the fallen oak, heavily, but with my penknife under my pillow. Twice in the night I woke and had to scramble out of the tent to go for a pee in the bushes, only to feel like I needed to go again as soon as I got back into the sleeping bag. The sound of the river tinkling nearby probably didn't help.

DAY FIVE

Slate slag heaps, a mad couple in Dolgellau, a non-vertigo-inducing toll-bridge and an unexpected visitor

IN the morning I went up to the toilet block, planning to have a shower, but once again I changed my mind when it came to removing my jumper. Perhaps I was enjoying the tramp lifestyle so much that I didn't want to spoil it all or it could have been the large, industrial-looking insect in the shower basin that put me off. Besides, even though it was early June, there was still a nip in the air in the mornings and evenings and the last thing I felt like doing was stripping off. What I really felt like doing this morning was lolling on a sofa supping gallons of hot tea. In my experience, when a campsite advertises hot and cold showers they usually mean literally that the showers go hot, then cold, then hot, before always finishing off cold.

In any case, there's something comfortable about your own personal grime: it acts as insulation, sun-block and as a real sign that you have transcended all thoughts of vanity and materialism. It may be the case that we only bother about personal hygiene because of the other people around us and, when there's no one looking, we soon stop caring about what we look and smell like. If I were stranded alone on a desert island, then looking presentable would certainly be right at the bottom of my list of priorities, that's for sure. While I was fully aware that I wasn't on a desert island, the lack of any real human contact had made me feel like I might as well be. So, still mucky and sporting a hairdo akin to that of Ken Dodd, I packed up my stuff and fired up the tractor, leaving the campsite behind me.

A lady trimming a hedge outside the front of her house confirmed that I could go north-ish towards Corris by following this road but that the main road was a more sensible option. I felt like explaining that a main road might be a sensible option if I were both in a car and in a rush but it wasn't worth trying to convert her to the delights of rural tractor driving so I thanked her and, ignoring her advice, continued along the wood-lined lane. In any case, she didn't look like the sort of person who was going to take lifestyle advice from a pregnant Ken Dodd lookalike on a vintage tractor.

Corris is a steep, little slate-built village with narrow streets, slate slab steps and a grey-green light to it. I drove aimlessly around for a few minutes, looking for someone to ask for directions and, while I was pulling over to have a think, I was overtaken by a VW caddy pick-up, being driven by a man in a flat cap. The man pulled up in front of me, got out of his pick-up and came over, smiling and saying 'Hi,' in a friendly and enthusiastic way.

For some reason, I immediately assumed that this bloke was someone I knew, though I couldn't for the life of me think who he was and why I was bumping into my mystery friend out in Corris of all places. I decided to go along with it all for the moment, hoping that the identity of this person would, at any second, come to me. So I said 'Hi,' back, in an equally friendly way and 'How are you!' – when what I really wanted to ask was '*who* are you?'

I said 'Yes, fine, great. . . .' when he asked me how it was going. Then he asked me where I was heading to and told me that he had a tractor exactly like mine, at which point I realised that I didn't know him at all.

He had obviously just stopped to talk to me because he was hugely interested in my tractor. I wondered if he thought I was mad, calling out 'Hi, how are you' like that but thankfully he didn't seem too fazed. He knew without being told that my tractor was a four-cylinder model and he told me that he was just on his way to do a bit of ploughing with his 35. 'You ever ploughed before?' he asked. 'No,' I replied.

Suddenly, I realised that I was in the presence of a tractor expert and, remembering the leaky fuel pipe, I said, 'Oh, before you go, you wouldn't have a look at this for me would you?' After all, if he happened to break the pipe, then he would, hopefully, feel obliged to mend it again for me. He looked at it, confirmed that it wasn't anything much to worry about and said that if he had had his tools with him he would have been able to tighten up the joint for me. I opened up my toolbox with a flourish and was able to find him the spanners he needed. I joked that I was one of those 'all-the-gear, no-idea' people and he laughed, saying that it was better to have 'all the gear, no idea' rather than 'no gear and no idea either', which I suppose is true.

Still, I couldn't help feeling like a daft bimbo who didn't know how to open the bonnet of her own car. He didn't seem to mind and I guess that like most men, he was enjoying a rare chance to be a hero. He said how marvellous it must be to be on a tractor trip like this, and in a fit of friendliness I found myself saying, 'Why don't you get your tractor and join me for a while?' He said, no, no, he had better get on with his work, but why didn't I come and try ploughing? I thought about it for a moment, then decided that no, no, I, too, had better get on with the job in hand. And since I still didn't know where I was going and neither did he, I had better find someone who did know. So I went on my way, thankful that knights of the road do still exist.

Five minutes later it dawned on me that the man might have thought I was coming on to him, shouting an over-enthusiastic greeting like that and then trying to get him to drive off into the sunset on his tractor with me. Oh no. It seemed I had lost the art of normal human communication, and it had happened so quickly. I could only hope that he hadn't gone and told his mates that there was a tractor-driving pregnant nymphomaniac at large in the countryside.

The edge of my next map showed a narrow, over-the-mountain lane, heading in the direction of Dolgellau; I just needed to find the start to this lane because that part, irritatingly enough, wasn't on the map. A postman in a van overtook me and when I saw him stop at a house further down the road, I roared after him on my tractor in order to ask him the way. I was certain that he would be a mine of information when it came to roads. Like the lady earlier, he told me that the main road was my best bet and I would be better off not starting from here. He also said that the lane I was looking for was really steep with many gates; in other words, it wasn't very nice to use.

It was as though people were somehow afraid of the wilderness, or else they thought of it as an unsuitable place for girls with tractors to go, so this time I explained how and why I wanted the remote and scenic roads. His directions went over my head, so I asked him if he would draw me a map and he waited patiently while I took everything out of the saddlebags in order to find my biro. I handed him a Llangrannog postcard so that he could draw me a map on the back. He glanced at the picture, raised his eyebrows and then drew me a map and a large arrow with 'here' written next to it, indicating the spot where he was standing. Then he handed me the drawing, said, 'Good luck then', as though I was planning to climb Everest, and we parted.

I began to wonder if this mountain road was dangerous for some reason, perhaps it was lethally steep and the gates were on such slopes that my tractor might fail to hold while I got off to open them. I decided that since the postman had confirmed that it was a tarmac road, then it certainly couldn't be any worse than my previous off-road-front-wheels-in-the-air experience, and I'd survived that.

The road was steep – cutting up and above the village, between the slate slag heaps and the seeded birch trees, higher and higher up into hilly farmland. The fact that there were several gates to open and close obviously put most people off using the road because I didn't meet a soul or a car, for that matter. There was nothing around, except some small, grubby mountain sheep, wobbly walls and fences that had seen better days. Gates are a great thing on country roads: they serve to deter busybodies and those people who are always in a hurry, which makes the road much more pleasant for everyone else.

When I dropped down the other side of the hill, I could see farms and homes dotted about on the greener plain below. The map indicated that I had to cross the main road in order to get to another green lane, which would take me in the vague direction of Dolgellau. I found this lane easily and, while I was closing the gate, I became aware that a farmer was standing and watching me from his drive on my right so I waved. He waved back with both arms, in the manner of someone trying to flag down an aeroplane. I guessed that for some reason he didn't want me to go that way, either that or double-arm waving was the custom in this area.

I left the tractor where it was and walked over to his gate. 'Where are you trying to go?' he asked me in Welsh. When I explained that I was about to go along the track, he told me that it was unlikely that I'd be able get through that way because in one place the track went as narrow as a footpath. There was a big boulder that jutted out and it was really only suitable for walkers. I told him that the OS map described it as being passable to vehicles and he raised his eyebrows towards the heavens and tutted, as if to say, 'What do they know, these people sitting in an office hundreds of miles away drawing maps. . . .'

I was unsure whether to take what he was saying as fact or whether he was saying it because he wanted to discourage off-roading types – like me – from using the track. Part of me wanted to at least have a go at getting around the boulder. But it seemed rude to ignore his advice, especially while he was still watching, so I turned the tractor around and drove a quarter of a mile further down the main road until I came to a small tarmac lane, which the map told me also went in the direction of Dolgellau.

When I reached Dolgellau, I decided I would park up in the town and stretch my legs. There was a convenient parking place right outside the shops and as I was tying down the ill-fitting old tarpaulin that covered my belongings in the back box, an elderly, rather wobbly, well-spoken English chap wandered over to talk to me. He told me he had been a tractor owner once and we talked for a while about the glories and the pitfalls of old tractors.

All at once, an elderly lady appeared at my side and said in a loud voice, 'Are you trying to steal my husband?' I wondered if the man in the pick-up had warned her about people like me and I apologised, saying that no, no, one man in my life was quite enough, thanks. She explained that her rather doddery husband was a real 'one for the ladies', and she had to 'watch him every minute'. They were quite a double act these two: as she was talking about him, he kept trying to deny it, saying that he was only trying to be friendly, but she wasn't letting him get a word in.

I thought that any minute she would start batting him around the head with her handbag and I couldn't help laughing, even though they both seemed deadly serious. Soon another lady joined us, saying, 'What's going on here then?' to which the first lady replied, 'This one is trying to get him into bed,' in a loud voice. By now people on the street were looking over so I attempted to smile in a way that I hoped said, 'Don't worry, they are all mad.'

I made my excuses and quickly wandered up the street, thinking that I must share my late father's talent for attracting peculiar people wherever I went. Partly to check that I wasn't being followed, I stepped into a tourist information shop and gazed at the large map they had on the wall, which showed the whole region I was in.

A polite tie-wearing man came over and asked if he could help me. I asked him what route was the most scenic one that I could take north and also if he was aware of any off-roading that I might be able to partake in while on my tractor journey. He was helpful, insofar as he showed me a long, high, single-track road which headed towards Trawsfynnydd, but he couldn't recommend any particular off-roading sections. He was keen to point out that driving across private land is a big no, no and I had to assure him that I wasn't the kind of person who did that kind of thing.

After that useful exchange, I was on my way up the street, window-shopping because I couldn't think of anything that I needed or wanted to buy. I gazed at some postcards, seeing a load of interesting places that I had passed but not visited, including an ancient, standing stone, which was now miles behind me. I kicked myself for not having researched the whereabouts of such things before my trip. I would never make it as a travel writer, I told myself. I had no agenda, I was just wandering and I was completely missing out on seeing some really important places. What a waster I was!

Soon I was heading out of the town, roaring up into the hills on a single-track road, which made me feel wonderful again. As is typical of the area, the lane was initially dotted with houses and cottages but as I climbed further away from the town, the roadsides became more deserted and soon I found myself on the high ground, which was inhabited by wiry grey-brown grasses and boulders instead of fields. I saw a big lonely pine tree on the side of the road and I remembered having read somewhere that pines once used to mark the places where drovers could leave their livestock for the night; subsequently, I wondered if this tree was the sole survivor of those times. It seemed a sad remnant of a bygone age and I felt for it: cold, alone and ageing on this desolate plateau.

While I was daydreaming about what it would be like to be a drover, I went over a cattle grid far too quickly and gave myself, the baby and the tractor a right good rattle. I stopped the tractor and lifted myself gingerly out of the seat with my hand over my belly to check for damage. I was fine. While a vintage tractor is utterly devoid of shock absorbers it seems that the human womb is specially designed to cope with bumps. From now on, I vowed to look where I was going instead of gazing over the colourless expanse of highland, day-dreaming about being a drover on horseback – complete with nice leather saddlebags.

The moment I began dropping down into the lower land, the landscape became green again. I came to intricate stone walls, paddocks with slate fences and low stone cottages – the kind of cottages that have privies up the back garden. I saw humans again but in nice, small quantities and in cute, yet awe-inspiring, surroundings. Far in front of me were the mountains – nothing like the rounded hills that I'd just come over – real, sharp up-and-down mountains, anthracite black, with their heads lost in an atmospheric smokescreen. The huddle of stone cottages, the lake in front, and the backdrop of the moody hills, made for a perfect image of rural Welsh life.

Then, out in front of me I saw it – the ominous and oddly incongruous image of a nuclear power station, looming over the quaint little village. My first reaction was anger – why did they build that here – in what must be one of the most beautiful landscapes in Britain? I felt like driving right in there and giving them what for. Of course I had seen this power station before, from the main road, and it was bad enough then, but up here, looking down on it from this most godly vantage point, it seemed so much worse.

I tried to take a photograph of the village, the lake and the mountains. Because I didn't want the power station in the scene, I positioned myself so that the building in

113

question was hidden by the exhaust (or chimney as I prefer to call it) of the tractor. But there was no getting away from the fact that it really was there. I doubted that anyone would ever have got away with building such a thing in somewhere like the Cotswolds, for instance. It isn't even as if anyone tried to make up for it a bit, by at least making the building itself attractive – it is morbidly functional – think 'glen of tranquillity', but with a giant, dubiously powered, concrete spaceship as the centre-piece. I guessed that it must have been built in this loca-tion because this was a poor area, one with a small enough population that wouldn't kick up a fuss, and might even be grateful for the work that such an industry might bring. But maybe I'm wrong about all that, maybe the locals were glad when the power station arrived, and maybe it's just me who shudders at the sight of it. I wondered though, how it was that planning per-mission had been given for it to be built there, when I know of a chap who is trying to do up a little ruined cottage that he owns, so that he can live there, and he isn't allowed to do so. Isn't life strange sometimes?

After passing the aforementioned monstrosity, I turned off the main road – yet again having neglected to put on the flashing beacon – and chugged off down a woody lane that headed towards the coast. I planned to try to meet the sea again, and to cross the estuary over to Penrhyndeudraeth by means of a toll-bridge. Having once been over the toll-bridge in a car, I was sure that this one was a nice, low, wide bridge, which would encourage my vertigo to remain safely in the closet. As I lane-pottered through the undulating landscape of woodland and meadows, I passed by entrances to houses; so idyllic it seemed that it was hard to remember that the nuclear power station was only about a mile away. Out of sight, out of mind.

As I drove along, I reflected on how my mind has wandered down all sorts of avenues of thought – far further than it does on regular work-and-home days – during this trip. I've found myself pondering over human behaviour, the nature of the universe and what it would be like to do this trip in another time or another era. This little excursion of mine was the closest thing to a spiritual retreat that I had ever experienced, and my brain had begun to react to the solitude. The everyday talking-and-doing part of my brain, it appeared, had become redundant and had occupied itself with fluffy philosophy and idle meanderings of the mind. Without the company of the wandering mind, I think that continuous tractor-driving could become quite boring.

Down another little lane leading to the toll-bridge, I passed a sign with a weight restriction on it. Having no idea what weight my tractor was, this sign meant nothing to me. As I passed, I told myself that big 4x4s must cross this bridge and surely my tractor wasn't heavier than one of them. I made a mental note to find out the weight of my tractor. After all, facts like that make it look like you know what you are doing. In an attempt to see what it felt like to have information like that at my fingertips, I said out aloud, 'It's a ton and a quarter un-laden, you know. . . .'

The gentle estuary and the long, low-slung bridge were all deserted. I took the bridge, feeling quite a buzz to be crossing something that afforded such excellent views of the winding waters and sand banks. Amazingly, there was someone sitting in a booth at the end of the bridge waiting to take the money. What a bizarre job, I thought. Once again, my mind wandered off with the question of what it must be like to work alone in that booth, looking over the bridge and the estuary below with only the occasional passing car to break the solitude. . . .

A while later, I found myself out in the little town of Penrhyndeudraeth, which seems like a long word because it is actually several words put together – namely the words 'two', 'beach' and 'headland', though not necessarily in that order. For a little and relatively unknown place, Penrhyn DD – as I affectionately call it – has quite a lot going for it. It has a quaint railway, complete with its own fat controller – last time I checked.

Also, Penrhyndeudraeth is home to Portmeirion, a life-size, Italianate-style village designed by Sir Clough Williams-Ellis. He proved that something large, man-made and functional, such as a village, can be artistically

116

shaped to fit in with, and even enhance, the surrounding area. Sadly, the idea never really caught on among architects and town planners but Portmeirion, with its rounded towers and painted archways, occupying a woody slope that curves around the estuary, remains there to be appreciated. Portmeirion is the strange place where the TV series *The Prisoner* was filmed. A local once told me that it was 'the worst and the weirdest place in which to take magic mushrooms'. I guess the place is peculiar enough as it is without the help of mind-bending hallucinogens.

As well as the cute railway and the fantasy village, Penrhyndeudraeth is also home to the Snowdonia National Park Offices. It isn't the most handsome of buildings but I've seen worse – like the power station, for instance. I suppose the National Park people are on a budget just like the rest of us, though it would have been setting a good example to us all if they had sensitively restored an existing building and used that as an office, rather than building a whole new one.

As I drove towards the outskirts of Penrhyndeudraeth, I looked over into a wall and saw a model village in someone's garden. I wondered if Sir Clough Williams-Ellis (deceased) wanted his full-sized model village to become the tourist venue that it is, or had he hoped it would end up being lived in by the public. Unable to decide which would have been the best outcome, I came along the main road that is The Cob, leading to the town of Porthmadog; or Portmadoc, as the English say.

Porthmadog is the port from which ships, loaded with slate from the surrounding quarries, once sailed to far-flung places like America. It is also the home town of Cob Records, which is the first record shop that I ever went into as a child accompanied by my father; also the place where I first realised that music was cool. While my dad scoured the rows of vinyl for second-hand Johnny Cash records, I loitered awkwardly, gazing with fascination at the punks, the hippies and the little-known rock stars of the Welsh music scene, as they browsed through the records, pulling on cigarettes and using the ashtrays that the management was able to provide at that time. It was a cosy, dingy place, where the love of sounds brought all sorts of people together. Apart from the fact that it now has CDs for sale, I don't think it has changed much at all – obviously smoking isn't permitted any more.

Apart from a local authority building, designed by some architect that needs a good slap round the head, Cob Records is the first place you come to after crossing the embankment. As I was looking fondly in at the car park of Cob Records, Father-to-be was stood there, waving at me. I pulled in, thinking how on earth? Of course I was pleased to see him but I also felt a bit thrown that he had just unexpectedly turned up. He only worked half an hour or so away from here and, after speaking to me this morning, he knew I would be passing Porthmadog sometime this afternoon. I could tell he thought he was giving me a nice surprise but the horrible, anti-social, controlling part of me had expected to see him tomorrow at the end of my trip and that would have felt like the perfect time for a joyous reunion. I found myself saying, 'I can't hang around here, you know. I've got to find a place to camp tonight.'

Then he told me, rather excitedly, that he had brought his sleeping bag with him and that he'd be able to join me on my last night. I have to admit that spending the evenings alone at the tent was becoming a little tedious but, all the same, this was going to be my last night and I had wanted to take my solitude right to the very end. So it was with mixed feelings that I found myself nodding and saying, 'Oh right.'

I looked at the map and suggested that we meet again a few miles further on at a campsite near Criccieth. I said that I would take a scenic detour in order to avoid the main road as far as possible. He asked if I wanted him to go in front and wait for me at various junctions along the way. It seemed pointless for him to join me in his van and I felt that being waited for at junctions would take away any feeling of adventure that I had left, so I declined the offer and he went on ahead to the campsite. This left me with the journey to the campsite to get over my annoyance at having my plans changed. As I drove I counselled myself about the perils of being a control freak. I told myself to get over it and stop wanting things to be 'just so' all the time.

By the time I got to the campsite I had recovered from the shock of having to change my plans and was looking forward to sharing a brew and a tent with Father-to-be and being able to tell him all about my past few days. He had sensibly found a quiet space near a hedge and together we put up the tent and got a brew on. I found that my time alone had definitely turned me into a worse control freak than normal: I couldn't stop myself insisting that the stove sat in its usual place on the back box of the tractor and that the tent was situated, not where he wanted it, but in a protected place between the tractor and the hedge. I had become tent-wise-tractor-travelling-survival-expert all of a sudden.

I have to say that having another person in my little camp felt peculiar. I felt the same way that a hermit must feel when a guest turns up out of the blue. It was as though I'd forgotten the art of basic human interaction and had turned into a solo-camping tractor freak. It wasn't so much that I didn't want to get back to real life – in fact I couldn't wait to see and feel home again – it was more that I couldn't talk about anything except camping and tractor driving.

For Father-to-be it had just been a few regular days but, for me, life had changed a bit. I'd been on a real adventure and I couldn't stop wittering on about it. I showed him my route on the maps; told him about driving up the side of a mountain; explained the Edwards brothers; the Museum of Power; the wonderful twisty lanes; and, most importantly, how it had felt to be there.

He listened politely and then said he thought that I'd been ever so brave. I didn't want him to think I was brave; there had been nothing brave about it. Any fool can drive a vehicle around some lanes and put a tent up. I just wanted him to understand what a strangely liberating and fascinating experience it had all been. I still don't think he quite saw the point in doing a trip like mine because if it had appealed to him, he would no doubt have done it himself before now. Still, I couldn't stop myself from selling the experience all the same. He probably felt like a someone having to hear all about someone else's brilliant holiday, painfully aware that in another week the photos will be back and he'll have to go through it all again in glorious Technicolor. I could just see myself describing the photos – 'Here's the tractor near the Nant y Moch reservoir,' and 'Here's the tractor relaxing by the tent on the third evening . . . doesn't it look sweet?'

While we could easily have jumped into the van that Father-to-be had met me in and gone off to get some tasty food at a pub, I felt the need to cook him exactly what I had been eating for the last few days – namely egg noodles – just so that he understood my experience. As we sat outside the tent eating, a couple with a baby pulled up next to us and started setting up camp.

'F---ing typical!' I muttered. 'As soon as you find a nice, isolated spot, someone pulls up alongside you . . . and if that isn't bad enough, they've got a baby with them.' Father-to-be patiently reminded me that soon we would be in possession of a baby and that it was interesting to note that it was still possible to go on camping trips with one in tow. I made a 'hmmm' noise. Then we said hello to the couple and asked them what it was like to take a baby camping. 'This is the first time we've tried it,' they said, 'so ask us in the morning!'

We wished them luck, silently hoping that we wouldn't be kept awake all night by incessant wailing from a baby who doesn't want to be in a tent. When it came to our bedtime I discovered that Father-to-be had decided that whatever I might wish to endure, he had no desire whatsoever to sleep on a thin piece of foam. He had brought a van full of duvets with him and set about using them to make a soft, squidgy nest on the floor of the tent. Feeling somehow proud of the hardships that I had endured, I was left with no choice but to call him a pansy for bringing all this stuff. But since the soft bedding was already here, I decided I might as well make the most of it. I still hadn't taken a shower. They were just too public; in any case, I was comfortable with my scruffiness and had decided long ago that the dirt was what was keeping me warm. Poor Father-to-be must have been shocked that, in just five days, his lady had turned into a tramp.

DAY SIX

*Criccieth, a creepy old house, a Cornish-style
coastline and feeling very, very cold*

IN the morning, Father-to-be woke complaining of a
bad back because of the unyielding field. 'Just
imagine what it was like without three bloody duvets
underneath you,' I retorted. After breakfast Father-to-
be kissed me goodbye, climbed into in his slightly
ridiculous little vintage van and left for work – it felt
quite strange waving him off from the tent, saying,
'Have a good day, see you later!' I felt like a suburban
housewife saying, 'Have a nice day at the office, dear,'
except that I was standing outside a tent, wearing a
nightie and a pair of unlaced steel toecaps. I wondered
what the couple with the baby, who were watching us,
must think.

When I went over to say good morning to them, it
turned out that the baby had kept them up all night and
they were too tired to care about anything: safe to say
that the sight of a pregnant woman wandering the
campsite in steel toecaps and a nightie hadn't even

registered. It crossed my mind, as of course it often did, that in just a few weeks I would be a mother and sleepless nights like that would be soon be mine.

Sleepless nights aside, the couple seemed happy enough, as though their bundle of joy more than made up for any hardships they had to endure. I hoped that I would feel like that once my baby arrived. It wasn't that I didn't like babies and toddlers, just that I never quite knew what to say to them. In some ways, I guess that I felt slightly unnerved by them and they seemed to know it. Horses are the same – they are always happier with someone confident – and I was sure that if I could soothe half a ton of horse, then I could handle an eight-pound baby. It would just take a bit of getting used to, that's all.

This was to be my last day on the road. It wasn't going to be as much of an adventure as the previous days simply because I spent five years living on the Llyn Peninsula, so much of it was familiar to me. The area would always have a special place in my heart: it was where I bought my first house, a beautiful and elderly – if rather crumbly and draughty – smallholding. And it was while I'd been living here that I'd bought my tractor. The tractor had spent its life on the peninsula, owned from new by one chap until he died. I bought it from his widow.

I can remember my late father trying to talk me out of the idea of buying a tractor. Over the years he'd seen a few accidents with farm machinery and he had it in his mind that I would manage to kill myself. He seemed to get over that initial fear after he came down to see the tractor and was pleasantly surprised to see that I hadn't gone and bought myself a complete wreck. Then I decided to buy a baler, so that I could make some hay bales for my horse and some to sell. Once again I was treated to a lecture on the '101 Ways to Kill Yourself with Farm Machinery'. I had to assure him that I definitely wouldn't be putting my fingers into the mowing machine while it was switched on (even if it did get clogged with a load of hay). I also wouldn't be driving the tractor while drunk – something that I knew he had done on at least one occasion – and I wouldn't run backwards kicking hay into the baler while it was going along – something my older brother used to do. Nor would I hang about by a working power take-off shaft while wearing flared trousers since they might get caught up and I'd be ripped to bits. Again, this was something that my older brother had once done, back in the seventies when flares were in – and he had almost been killed in the process.

Yes, I promised him that I would be sensible and that I had learned from other people's mistakes. What he didn't realise was that I was actually quite nervous about the tractor and, because of that, I was hardly likely to get cocky (as he put it) with my new machine. The noise that the power take-off – the shaft at the back of the tractor which powers implements like hay turners and balers – made when it was in operation was quite alarming to a newcomer, as were the vicious-looking spikes with which all haymaking implements seemed to be armed.

He was still unhappy about me making hay with all of this potentially lethal equipment. We were speaking on the telephone once when I said in an exasperated tone, 'I'm not completely bloody stupid you know!' The phone call ended with him shouting, 'Well I wash my hands of it all then!'

The very next day was the day I'd planned to give my £60, rust-bucket of a baler a first go and, despite the conversation, he just happened to arrive in the yard on his motorbike. 'I didn't know that you were coming,' I said and, though he lived almost sixty miles away, he told me he was just passing. At the time, I felt a bit insulted that he thought I was just a stupid girl and that my lack of a penis made him assume that I'd be utterly incapable of operating farm machinery. With hindsight, I can see that he'd not long lost my mother and he was probably afraid that he'd lose me too. Shouting about what a stupid idea it was to buy a tractor was his way of showing that he loved me.

Now back to my journey. Since I was familiar with the area, I knew that I would easily get to Whistling Sands by the afternoon, even by taking the long scenic route. From there, my tractor and I would be picked up by Father-to-be and that would be the end of it all. I had mixed feelings about being on the last leg of the journey. On the one hand, the novelty of solo camping was beginning to fade and I missed my dog and my bed – and Father-to-be a bit too. On the other hand, I was sad to be leaving my little adventure – it all seemed to have passed so quickly – and real life was now beckoning me home, saying, 'Like it or not, it's time to get back to normal now. Time to face your responsibilities.'

I tried to remember how I used to spend my days. Going to work at an undertaker's office in an immaculate black suit felt like a lifetime back. If only the firm could see me now, all scruffed up with my hair standing on end – they would never have given me the job in the first place.

I packed up my stuff for the last time, fired up my 35 and left the campsite. It was another bright day and I'm sure that behind glass it would have been pleasantly warm but from the exposed seat of my tractor a thin sea breeze nipped at my collar, dictating that I wore all at once almost all the clothes I had with me. I was acutely aware of how stupid a pregnant woman wearing a balaclava on a tractor must have looked, so as I approached the town of Criccieth I pulled the garment down around my neck and placed another hat on top. I tore through the main street of Criccieth, noting that it was a pleasant little town with a beach and a crumbling, yet commanding, castle and lots of tall and brightly painted bed & breakfasts. I also passed an interesting café out on the seafront that was designed by none other than Sir Clough Williams-Ellis – of Portmeirion fame – to be a kind of ballroom, instead of a café.

Now, however, it served ice creams and fry-ups and had a refrigerated display cabinet at the counter, displaying rows of macabre-looking dolls. It was an attractive, airy building and it was great that it was still in use, even if it was full of sunburned, burger-eating tourists, rather than the glamorous dancing couples that it was meant to entertain. I was particularly amused by the sight of a large, old-fashioned weighing machine in the café doorway. This machine was like the sort that used to be outside the chemist shops of my childhood, where you put five pence in and stood on to weigh yourself, out on the street, for all to see!

After Criccieth, I had a piece of uninspiring main road to endure before I could take to the lanes once again. I was looking for a dirt track called Lon Goed, which would take me through a wood and further onto the peninsula. Lon Goed turned out to be a nice track through a beech wood but was also one of those places where dogs needing to relieve their bowels get taken for walkies, where people dump fridges and piles of rubble and where shifty-looking couples sit in cars. During the course of my travels along Lon Goed, I passed a few furtive 'couples', several straining dogs with their sheepish-looking owners and one fridge lying on its side with its mouth wide open. Then I came back out onto tarmac. This is the trouble with tracks like this, you drive miles in order to locate one, often to find that it only takes for you a mile – if you're lucky – and then you find yourself back on a road again, usually somewhere you hadn't particularly wanted to be.

I realised that already I was lost. The map showed a junction where there wasn't one in real life and, with that, I decided that I might as well abandon the paperwork altogether and just head west. The Llyn Peninsula isn't all that wide and as long as I was somewhere on it, heading towards the end, it really didn't matter what road I was on. Since I used to live near the end of the peninsula, I knew that area would be familiar but this bit, the upper arm of the peninsula, was a maze of little lanes that were new to me. It made a pleasant change not to have to study the map continuously; there was a freedom in just being able to follow my sense of direction. Some lanes, however, seemed to start off going west but after they turned and twisted, you began to think that before long you would be back where you started.

Being high up on the tractor helped because, with the extra height, I could easily see the hills of Garn Fadryn and Mynydd Rhiw further down the peninsula. These landmarks were like beacons, showing me the way I needed to head. It would be nigh on impossible to use your sense of direction like this if you were in a car because the lanes are all flanked with high banks, some with hedges on top, and unless you were high up, it would be hard to see anything but the bit of road in front of you.

The Llyn Peninsula is dotted with houses and farms, each one a polite distance from the next, so while it may not feel totally remote, it's not crowded either. This is farming country and everyone drives tractors, though mostly they are a lot more modern than mine. While the tourists come for the beaches when the weather is right, the winding back lanes are generally used only by locals, and since there aren't that many locals, the single-track roads are pretty quiet.

I came to a shady lane, where the trees hung low over the road, almost meeting in the middle, making the place dark and tunnel-like; in stark contrast to the otherwise wide open countryside and big sky. Alongside this shadowy lane was a high wall, which looked like it was the boundary of an estate or manor house of some kind. Before long I passed what must have been the main entrance to the house, the antique-looking gates were locked, moss covered the drive and it didn't look as though anyone had been there for years. Through the overgrown rhododendrons I could see the upper part of the house. It was tall and rambling and although it was by no means derelict, it had a sense of dark abandonment about it. *Great Expectations* and the character of Miss Havisham immediately came to my mind and I wondered if this place really was empty, or if it was occupied by an elderly recluse, who for one reason or another had allowed the evergreens to close in around them, shutting out the bright Llyn sunshine.

I imagined grand staircases covered in cobwebs, dark oil paintings of grim-looking relatives hung in the silent corridors and the moss and ivy gradually taking hold. . . . My imagination told me that there must be some curious story to this silent, old place and that thought sent a little shiver down my spine.

Later, back under the big blue sky, I rounded the windswept shoulder of Garn Fadryn, which is a tiny but perfectly formed mountain and came face-to-face with a guy in a Land Rover who expertly reversed back for me, which told me straight away that he was a local. As he went to pass me I realised that I knew this chap from my time of living down here. He wound down his window and said, 'Hey!' in surprise.

Being the sort of guy who always asks what you are doing and then never gives anything away about what he is doing, he duly asked me what I was up to. I told him proudly that I had just driven the tractor up from Pendine Sands, and he refused point blank to believe me. He kept saying, 'Stop mucking about, what are you really doing here?' I protested that I really was at the tail-end of an epic journey. He wouldn't have any of it and bade me farewell, winding up his window and shaking his head as he went on up the mountain lane.

It had amused me, when I lived down here, that people called these little hills mountains and I always wanted to say, 'That's not a mountain – Snowdon is a mountain!' Which is similar to the way that people from the Himalayas might say, 'Snowdon – that's not a mountain, these are mountains!'

Maybe it was the fact that I had been on my tractor for too long, or maybe it was the fact that I was now in an area I knew well; either way, I couldn't seem to pluck up much enthusiasm for my beautiful surroundings. I felt tired and cold, and because I had seen so much awe-inspiring scenery on my trip, I had become used to it and was starting to take it all for granted. 'Another spectacular view of blue sea and craggy coastline . . . Oh, and another quaint farmhouse with a sheepdog outside. . . .' I really had been spoiled and I had to keep reminding myself that I was in an area of outstanding natural beauty.

For the first time in my trip I began to wish that I was travelling by car, only because, though it was a sunny day, my body felt like it was shutting down from the cold. I had to keep looking at my mobile phone to see if I had a signal on it; I had tried to call Father-to-be to say that I was nearing my destination so he might as well set out with the trailer to meet me, but I hadn't anticipated that much of the Llyn Peninsula would be a mobile phone black-spot. By now I wasn't far from Whistling Sands which meant that, while I would have the joy of having arrived, I would also have a long wait for Father-to-be, since he was at least two hours' drive away. Really I should have gone to a village and sat in a café or a pub for part of my wait but my tiredness made me want to keep away from any potentially social situations. With that in mind, I carried on in the direction of my final beach, and before long I came to a place where there was a signal. I phoned Father-to-be and told him that I was almost at my destination.

I chugged slowly between the hedges and the rolling fields, able to see the sea on three sides most of the time. When there is nothing to do but look at hedges, fences and fields, you soon begin to take a huge interest in otherwise mundane things. I would note where someone had done a fine job of repairing a stone wall or where the hedge had been neatly laid. I would tut over untidy boundaries and unimaginatively built breeze block walls. With nothing better to do, I had become a member of the architecture police, scouring the countryside for ugly buildings, eyesores and blots on the landscape. Some people obviously went to huge efforts to build their places in keeping with the quaintness of the area; others, it seemed, didn't care at all about the fact that their huge, concrete bungalows looked completely at odds with the gentle rolling fields and the old stone gateposts.

Whenever you get anywhere near the sea in Britain, you are sure to see a pet hate of mine, namely air bricks

– those far-too-uniform-looking concrete blocks with shapes cut out of them. They are the seaside town-person's answer to the picket fence and while they may once have been modern, they now look dated and taste-less – in my opinion.

So along I chugged, giving certain houses and bound-aries the thumbs up and others the thumbs down. When it was a thumbs down I made a sound – exactly like the one that you hear on *Family Fortunes* when someone has given a wrong answer – out loud. Sadly, I found that most of the newer houses were ugly, which must mean that either we are going through a particularly awful phase in the fashion of home design, or worse, that we are no longer able to make nice houses. Despite the awfulness, it was a most enjoyable way to pass the time, with the height of the tractor giving me an excellent viewpoint from which to pass judgement on almost every home that I passed.

On a sharp corner, I passed a tall, rangy-looking Scots pine in a field. It had an arthritic-looking arm out-stretched, pointing in the direction of Whistling Sands and it stood out like the outsider it was, amongst the low hawthorn hedges and elderflower scrags. Sure enough, just up the road was the turning to the beach of Whistling Sands. I drove right down to the beach for a look. The waves were rolling white, the wind was up and a few hunched souls walked aimlessly along the beach with their hands in their pockets. It was a bit of an anticlimax really, probably because I just felt so cold and tired.

I turned around, drove back up to the car park at the top and pulled the stop button. I should have got a celebratory cup of tea on the go, or at least gone for a walk to get warm, but I had no go left in me. It was all I could do to get off the tractor and sit down on the floor with my spine resting easily against the back wheel. I pulled my coat around me and huddled, which resulted in me almost, but not quite, drifting into a deep, dribbly sleep, all the time getting colder and colder. I kept thinking, Get up! Walk around! But I couldn't muster any strength, so instead, I periodically took my phone out of my pocket and gazed at it expectantly, despite the fact that it didn't have any signal.

I wasn't sure if I had nodded off or not but it definitely seemed like I had been waiting for a very long time and Father-to-be still hadn't arrived. If something was wrong, there was no way he could phone me and I wondered how long I should wait before leaving to find a place with signal. I realised then that only an hour had passed and it would take him longer than that to get to Whistling Sands, so I pulled up my knees, rested my head on them, closed my eyes and continued to huddle against the back wheel, fully in the throes of one of those 'ooh I've come over all peculiar and tired' pregnancy moments.

After what seemed like an age, I heard the rattle of a

Land Rover and looked up to see Father-to-be turning around in the car park. I stood up, feeling stiff, pale and shivery and waved feebly. But before he would load the tractor onto the trailer, Father-to-be wanted to take a photo of me and my tractor down on the beach. I felt weak and perished as I posed, smiling just long enough for the shutter to go down. We loaded the tractor and jumped into the Land Rover, turning the heater on full blast – despite knowing that the heating function on a Land Rover tends to make a loud whirring noise without actually giving forth any real heat. After half an hour under my sleeping bag, I began to thaw and come back to life.

Over the roar of the 'all-mouth-no-trousers' Land Rover heater, we discussed my little adventure. Father-to-be said, 'You'll have to do another one some day. . . .' And I said I didn't think that it was likely, not for a long time at least. To which he replied, 'Well I can look after the baby, I mean, I'm not suggesting you go the week after it's born but, you know, once life has got back to normal.'

I had to admire his boundless optimism and his utter belief that life would actually return to normal after having a baby. I smiled to think that he might actually think that I was the sort of person who would come straight out of hospital, hand him a baby and say, 'I'm off on the tractor now, see you in a couple of weeks.'

Between the two of us, we made one quite-well-balanced person. Me with my grimly realistic – or negative depending on your perspective – views on life and him with his happy, head-in-the-sand optimism. It was great that he fully believed in himself enough to know that he would be capable of taking care of a baby, as any father should, of course.

We can all learn things, if we want to. Just the same as I can learn how to look after a baby. . . . There were going to be plenty of people to ask for advice, which is partly why I had already decided that I would have the baby in hospital, if it gave me the chance to get there. At least then there would be people around who knew all about babies and I wouldn't leave until I knew everything I needed to know. After all, no one buys a crocodile without reading up on it first, not unless they're a complete lunatic, that is.

Back home, my friend came down with a bottle of champagne. I had a drink despite the pregnancy – because, as the old people say, 'it's already cooked by now anyway' – and one glassful left me high as a kite. I was being treated like a Grand Prix winner, and I felt rather like a fraud. I stood by what I had thought earlier, I had just driven along some lanes, which was something anyone could do. All the same, it was pretty touching.

SIX WEEKS LATER

I T was the day the baby was due to be born. By now I developed a stitch whenever I went for a walk and, though I was keeping as active as possible and had suffered from none of the really awful side effects of pregnancy, I had begun to take things easy. You might think from reading all about my tractor trip that somehow I had Father-to-be under the thumb, getting him to pick me up and drop me here and there, but that certainly isn't true.

For instance, for the final six weeks I was made to sleep on a bed with a big black silage bag underneath the sheet. All right, it was an unused one but all the same it was sweaty, it crinkled loudly whenever I moved and its slipperiness caused the sheet and under-blanket to slide off in the most irritating way. All of this is because Father-to-be spoke to women who had their water dramatically burst and make a right mess of the soft furnishings. He had recently bought a new mattress and he couldn't cope with the idea that he might have to shell out for a new one if my water burst while I was in bed. In any event, that scenario never did happen so I was forced to sleep, week after week, on that infernal silage bag for nothing.

As time went on I got more and more waves of tiredness and I used to have to pop to the doctor's surgery every Thursday to see the midwife. One day I went in and said that I felt shattered, like I had lead in my boots. Since we had been concreting a path that day I still had my steel toecaps on and the midwife looked down and said, 'Well it looks to me like you have got lead in your boots.' She was just that kind of woman – the reassuringly dry kind – and she hadn't quite forgiven me for coming back saying that the tractor trip had done me the power of good.

Anyway, came the day when the baby was supposed to arrive. I had learnt that, although this was my due date, apparently there was little likelihood of the baby actually being born on it, which meant that the date was more of a guideline than anything else. We were outside and I think I was emptying the ash pan or doing something behind the car, while Father-to-be was nearby, deeply involved in a conversation with my brother, who was further down the yard. They were talking about crankshafts and rotations per minute when suddenly I felt this almighty cramp-like pang. I held onto the wing of the car, moaned and took deep breaths, bent over. After a couple of minutes the pain faded and I looked up and saw that neither man had noticed: the engine-based conversation was flowing along, just as before. That's when I realised that I was going to have to make more noise if I wanted to get noticed.

139

Fifteen hours later, the pains became more frequent, harder to bear and somewhat noisier. By now Father-to-be was aware that something was happening. In fact, he was trying to sound and appear calm but his acting skills weren't up to much. Six o'clock the following morning suddenly seemed like the right time to leave for the hospital. I hadn't wanted to get there too early and be hanging about but, at the same time, I didn't relish the thought of giving birth in a Land Rover on the side of an expressway. At this point, I imagined that I was about two hours away from giving birth so leaving then would give me an hour in the hospital to settle in. Having a contraction in a Land Rover wasn't nice but then having a contraction anywhere isn't nice, oddly enough, and there are probably worse places to have them.

Turns out, I was mistaken about the timing. It was almost twenty-four hours later that I actually had the baby. This wasn't because of complications: it was all just happening very, very slowly. It happened so slowly, in fact, that people changed shifts and then the first lot of people all came back, then left again, and here I was, in too much pain to sleep, but not actually having the baby either. On TV, people seem to have babies within half an hour but I'd been at the hospital so long that I'd started to get to know all about the lives of the staff.

Finally things accelerated and my last real memory of the outside world was yelling at Father-to-be to do something about the bloody radio – it was playing off-station opera, which is the last thing that anyone, even an opera fan, needs as a birthing soundtrack. The midwife seemed to think I should be paying more attention to the birth and less to the accompanying music, but by now the laughing gas, the pain and the exhaustion were making me forget what it was I was here for. At long last, a baby girl arrived.

You will be relieved to know that I didn't call her Massey, or Fergie, or anything awful like that.

ABOUT THE AUTHOR

Josephine Roberts has a degree in philosophy and lives by the ethos 'Make do and Mend' whenever possible. She is a regular contributor to *Tractor* magazine and the American publication *Farm Collector*. In her free time she is passionate about the environment and committed to the preservation of rural heritage as well as Welsh culture.

A vintage tractor enthusiast, she owns a little red 1960 4-cylinder diesel Massey Ferguson 35 which she uses to maintain her smallholding in the foothills of Snowdonia. Josephine lives with her two young children and her partner in a house that has been in her family for three generations.

Extract from the author's travel diary

Other Books from Old Pond Publishing

Charismatic Cows and Beefcake Bulls   SONIA KURTA
Sonia Kurta's memories of farm work as a young girl are mostly set in Cornwall on the great Caerhays estate. She joined the Land Army in 1943 and stayed until it was disbanded in 1950.

Footsteps in the Furrow   ANDREW ARBUCKLE
The author presents a picture of farming life in Fife over the last century to show what it was like for those who worked in the industry. In 2007 he began 'harvesting' the memories of the older generation; their stories are often amusing or describe long-gone farming methods.

A Good Living   HUGH BARRETT
Managing farms from 1938 to 1949, Hugh Barrett encountered a range of characters from gentlemen farmers to ex-miners on land settlement holdings, with wartime profiteers and downright rogues for good measure.

Land Girls Gang Up   PAT PETERS
The group of 17-year-old London girls sent to pick potatoes in Cornwall in 1942 created havoc wherever they went. They fought a running battle with the local farmers with humorous and long-lasting effects.

Secret Rivers   PAUL HEINEY FOR ITV ANGLIA
In these two television series Paul Heiney canoes down fourteen East Anglian rivers, meeting folk who live and work along their rich but little-known banks.

We Waved to the Baker   ANDREW ARBUCKLE
A series of childhood memories which are nostalgic, amusing and full of warmth. There were four children in this farming family, three boys and their younger sister, who often courted unintentional disaster at the hands of her brothers.

Free complete catalogue:

Old Pond Publishing Ltd
Dencora Business Centre
36 White House Road, Ipswich IP1 5LT, United Kingdom
Secure online ordering: **www.oldpond.com**
Phone: 01473 238200   Fax: 01473 238201